PELICAN BOOKS

# AN INTRODUCTION TO
# CONTEMPORARY HISTORY

Geoffrey Barraclough was born in 1908 and educated
at Bootham School, York, and Oriel College, Oxford.
In 1934, after four years spent in Rome and at the
University of Munich, he became a Fellow of Merton
College, Oxford, but subsequently moved to St
John's College, Cambridge. After the war – during
which he served in the R.A.F.V.R. – he held pro-
fessorships at Liverpool University and the London
School of Economics, before taking up a chair in the
University of California in 1965.

Amongst Professor Barraclough's many publica-
tions are: *Mediaeval Germany* (1938), *The Origins
of Modern Germany* (1946), *History in a Changing
World* (1955) and *European Unity in Thought and
Action* (1963). He is a regular contributor to the
*New Statesman*, the *Observer*, the *Guardian*, and
other journals; and he has lectured in the United
States, Canada, Poland, Holland, Italy and
Germany.

LIBRARY OF

МGМ

КАЗАНСКИЙ УНИВЕРСИТЕТ

# AN INTRODUCTION TO

# CONTEMPORARY HISTORY

*Geoffrey Barraclough*

PENGUIN BOOKS

Penguin Books Ltd, Harmondsworth, Middlesex, England
Penguin Books Inc., 7110 Ambassador Road, Baltimore, Maryland 21207, U.S.A.
Penguin Books Australia Ltd, Ringwood, Victoria, Australia

—

First published by C. A. Watts 1964
Published in Pelican Books 1967
Reprinted 1968, 1970 (twice)

—

Copyright © Geoffrey Barraclough, 1964, 1967

—

Made and printed in Great Britain by
Hazell Watson & Viney Ltd,
Aylesbury, Bucks
Set in Linotype Baskerville

This book is sold subject to the condition
that it shall not, by way of trade or otherwise,
be lent, re-sold, hired out, or otherwise circulated
without the publisher's prior consent in any form of
binding or cover other than that in which it is
published and without a similar condition
including this condition being imposed
on the subsequent purchaser

# CONTENTS

# PREFACE

THE theme of this book was first tentatively developed in a paper read to the Oxford Recent History Group in 1956. Circumstances during the subsequent five years prevented my working on it further, and I am greatly in debt of those whose help and encouragement enabled me to take it up again. In particular, I should like to express my gratitude to the Rockefeller Foundation for their generous support and to the Master and Fellows of St John's College, Cambridge, for their hospitality.

The basis of this book is the Charles Beard lectures delivered at Ruskin College, Oxford, in the spring of 1963 and (in a revised form) at the University of California, Los Angeles, a year later. The appearance of this Pelican edition has provided still another opportunity for revision, which I have gratefully used (particularly in Chapters I, V and VII) both to bring the text up to date and to introduce new fact and illustration.

In attempting to single out what seem to me to be some of the main themes of contemporary history, one of my purposes has been to clear the way for the narrative history of the world since 1900 which I have in preparation. It seemed to me that a theoretical framework, which attempted to clarify the basic ideas and place events in perspective, was an essential preliminary to any chronological survey. But the present book is, of course, complete in itself, and I hope that those who believe that it is important to explore the historical foundations of the contemporary world will find it useful and interesting.

G. BARRACLOUGH

*July,* 1966

# I

# THE NATURE OF CONTEMPORARY HISTORY

*Structural Change and Qualitative Difference*

WE live today in a world different, in almost all its basic preconditions, from the world in which Bismarck lived and died. How have these changes come about? What are the formative influences and qualitative differences which are the distinguishing marks of the contemporary era? It is with these questions that the present book is concerned, and for that reason I have called it an introduction to contemporary history. It is not an introduction in the familiar sense of providing an elementary narrative account of events in Europe and beyond Europe during the past sixty or seventy years. Merely to recount the course of events, even on a world-wide scale, is unlikely to result in a better understanding of the forces at play in the world today unless we are aware at the same time of the underlying structural changes. What we require first of all is a new framework and new terms of reference. It is these that the present book will seek to provide.

Our search will carry us along some unfamiliar, or less familiar, paths. Historians of the recent past have assumed for the most part that, if they explained the factors leading to the disintegration of the old world, they were automatically providing an explanation of how the new world emerged; and contemporary history has therefore consisted largely of accounts of the two world wars, the peace settlement of 1919, the rise of Fascism and National Socialism, and, since 1945, the conflict of the communist and the capitalist worlds. For reasons which will appear later, this line of approach seems to me inadequate, in some ways perhaps even misleading. We shall be concerned here far more with the new world coming to life than with the old world that was dying, and we only need to look around us to see that some of the most characteristic features of the contemporary world have their origins in movements and developments that took place far away from Europe. One of the distinctive facts about contemporary history is that it is world history and that the forces shaping it cannot be understood unless we are prepared to adopt world-wide perspectives; and this means not merely supplementing our conventional view of the recent past by adding a few chapters on extra-European affairs, but re-examining and revising the whole structure of assumptions and preconceptions on which that view is based. Precisely because American, African, Chinese, Indian and other branches of extra-European history cut into the past at a different angle, they cut across the traditional lines; and this very fact casts doubt on the adequacy of the old patterns and suggests the need for a new ground-plan.

It will be one of the main contentions of this book that contemporary history is different, in quality and content, from what we know as 'modern' history. Looking back from the vantage-point of the present, we can see that the years between 1890, when Bismarck withdrew from the political scene, and 1961, when Kennedy took up office as President of the United States, were a watershed between two ages. On one side lies the contemporary era, which is still at its beginning, on the other there stretches back the long vista of 'modern' history with its three familiar peaks, the Renaissance, the Enlightenment and the French Revolution. It is with this great divide between two ages in the history of mankind that this book will chiefly be concerned; for it was then that the forces took shape which have moulded the contemporary world.

1

It must be said immediately that many historians – perhaps a majority of historians at the present time – would question the validity of the distinction I have drawn between 'modern' and 'contemporary' history and would deny the existence of a 'great divide' between the two. For this there are a number of reasons. One is the vague, indefinite, almost nebulous character of the concept 'contemporary', as it is commonly used. Another, which is more fundamental, is the tendency of historical writing today to emphasize the element of continuity in history. For most historians contemporary history does not constitute a separate period with distinctive characteristics of its own; they regard it rather as the most recent phase of a continuous process and, chary of admitting that it is different in kind or quality from earlier history, treat it simply as that part of 'modern' history which is nearest to us in time.

It is unnecessary to enter into a lengthy discussion of the reasons why I find this attitude difficult to accept.[1] In my view continuity is by no means the most conspicuous feature of history. Bertrand Russell once said that 'the universe is all spots and jumps',[2] and the impression I have of history is much the same. At every great turning-point of the past we are confronted by the fortuitous and the unforeseen, the new, the dynamic and the revolutionary; at such times, as Herbert Butterfield once pointed out, the ordinary arguments of causality are 'by no means sufficient in themselves to explain the next stage of the story, the next turn of events'.[3] There is, in fact, little

1. They are briefly discussed in my book, *History in a Changing World* (Oxford, 1955), pp. 4 ff.

2. cf. Bertrand Russell, *The Scientific Outlook* (London, 1931), p. 98.

3. cf. H. Butterfield, *History and Human Relations* (London, 1951), p. 94.

difficulty in identifying moments when humanity swings out of its old paths on to a new plane, when it leaves the marked-out route and turns off in a new direction. One such time was the great social and intellectual upheaval at the turn of the eleventh and twelfth centuries which we so inadequately call the Investiture Contest; another, it is usually agreed, was the period of the Renaissance and Reformation. The first half of the twentieth century has all the marks of a similar period of revolutionary change and crisis. Here, again, we are led to one of the central problems in the writing of history – the problem of periodization – and it would take us too far to discuss the theoretical issues it raises. But if we view the fifty or sixty years beginning around 1890 from this standpoint, it is difficult to avoid certain important corollaries. The first is that the twentieth century cannot be regarded simply as a continuation of the nineteenth century, that 'recent' or 'contemporary' history is not merely the latter end of what we call 'modern history', the most recent phase of a period which, according to conventional divisions, began in western Europe with the Renaissance and the Reformation. And if this is true, it would seem to follow that the standards of measurement we apply to contemporary history should be different from those applied to earlier ages. What we should look out for as significant are the differences rather than the similarities, the elements of discontinuity rather than the elements of continuity. In short, contemporary history should be considered as a distinct period of time, with characteristics of its own which mark it off from the preceding period, in much the same way as what we call 'medieval history' is marked off – at any rate for most historians – from modern history.

If these propositions have any degree of validity, it would seem reasonable to conclude that one of the first tasks of historians concerned with recent history should be to establish its distinguishing features and its boundaries. In doing so we must, of course, beware of false categories (that applies to all historical work); we must remember that all sorts of things last over from one period to another, just as all sorts of things regarded as 'typically medieval' persisted into Elizabethan England; and we should not expect to assign fixed dates to changes which, in the last analysis, are only changes in balance and perspective. But it still remains true that unless we keep our eyes alert for what is new and different, we shall all too easily miss the essential – namely, the sense of living in a new period. Only when we have the real gulf between the two periods fixed in our minds can we start building bridges across it.

It goes without saying that we can only consider contemporary history in this way when we are clear what we mean by the term 'contemporary'. The study of contemporary history has undoubtedly suffered because of the vagueness of its content and the haziness of its limits. The word 'contemporary' inevitably means different things to different people; what is contemporary for me will not necessarily be contemporary for you. It is still possible to meet people who have conversed with Bismarck,[1] and (to mention but one personal recollection) my old colleague in Cambridge, G. G. Coulton, who died in 1947, was a schoolboy in France before the Franco-Prussian war, and still possessed his school uniform with *képi* and baggy *pantalon* trousers – a diminutive version of the uniform of the French infantryman of the day – which he got out of storage for my eldest son to try on.[2] On the other hand, there is already a generation in existence for which Hitler is just as much an historical figure as Napoleon or Julius Caesar. In short, 'contemporary' is a very elastic term, and to say – as is often done – that contemporary history is the history of the generation now living is an unsatisfactory definition

1. cf. Golo Mann, 'Bismarck and Our Times', *International Affairs*, vol. XXXVIII (1962), p. 3.
2. Coulton recounts his three terms in St Omer in *Fourscore Years* (Cambridge, 1943), pp. 39–47; it was in 1866–7.

for the simple reason that generations overlap. Furthermore, if contemporary history is regarded in this way, we are left with ever-changing boundaries and an ever-changing content, with a subject-matter that is in constant flux. For some people contemporary history starts in 1945, with perhaps a glance back to 1939; for others it is essentially the history of the inter-war years or, a little more widely, of the period from 1914 to 1945, and the years after 1945 belong to a phase which is not yet history. The German Institute of Contemporary History, for example, is concerned primarily with National Socialism, the origins of the National Socialist movement under the Weimar republic, and the resistance movements which National Socialism provoked,[1] and it is possible to find able and intelligent discussions of the practical problems of writing the history of contemporary events which ignore – clearly not accidentally – anything after the end of the Second World War.[2]

The problems involved not only in the writing but also in the conception of contemporary history have given rise, ever since 1918, to a long, contentious, and ultimately wearisome controversy.[3] The very notion of contemporary history, it has been maintained, is a contradiction in terms. Before we can adopt a historical view we must stand at a certain distance from the happenings we are investigating. It is hard enough at all times to 'disengage' ourselves and look at the past dispassionately and with the

1. cf. H. Rothfels, 'Zeitgeschichte als Aufgabe', *Vierteljahrshefte für Zeitgeschichte*, vol. I (1953), p. 8; the same attitude is adopted by B. Scheurig, *Einführung in die Zeitgeschichte* (Berlin, 1962), pp. 30–1.

2. cf. M. Bendiscioli, *Possibilità e limiti di una storia critica degli avvenimenti contemporanei* (Salerno, 1954).

3. It can be followed in the pages of the journal *History*, beginning with the controversy between E. Barker and A. F. Pollard in 1922 (vol. VII); there followed R. W. Seton-Watson's plea for the study of contemporary history (vol. XIV), renewed by G. B. Henderson in 1941 (vol. XXVI), and further contributions by David Thomson (vol. XXVII), Max Beloff (vol. XXX) and F. W. Pick (vol. XXXI).

critical eye of the historian. Is it possible at all in the case of events which bear so closely upon our own lives? It must be said immediately that I have no intention of entering into a discussion of these methodological questions.[1] Whatever may be the problems of writing contemporary history, the fact remains – as R. W. Seton-Watson long ago pointed out[2] – that, from the time of Thucydides onwards, much of the greatest history has been contemporary history. Indeed, if it is said – as historians sometimes say – that the idea of contemporary history is a newfangled notion introduced after 1918 to pander to the demands of a disillusioned public anxious to know what had gone wrong with the 'war to end all wars', it is not unfair to answer that what was newfangled was not a concept of history firmly anchored to the present but, on the contrary, the nineteenth-century notion of history as something dedicated entirely to the past. What is *zeitgebunden* – what, in other words, is a product of the identifiable circumstances of a particular time – is not the belief that contemporary events fall within the historian's ambit but the idea of history as an objective and scientific study of the past 'for its own sake'.[3]

On the other hand, it would be idle to deny that those who reject contemporary history on the ground that it is not a serious discipline are in practice frequently proved right. Much that claims to be contemporary history – whether written in Peking or Moscow, or in London or New York – turns out too often to be little more than propaganda or desultory comment on 'current affairs', reflecting usually an obsession with one aspect or another of the 'cold war'. The pitfalls to which such writing is liable are obvious. What prospect is there, for example, of

1. They are briefly reviewed by H. Rothfels, *Zeitgeschichtliche Betrachtungen* (Göttingen, 1959), pp. 12 ff.

2. *History*, vol. XIV (1929), p. 4.

3. This was demonstrated, with great verve and learning, in Fritz Ernst's brilliant article, 'Zeitgeschehen und Geschichtsschreibung', *Die Welt als Geschichte*, vol. XVII (1957), pp. 137–89.

assessing realistically the Castro revolution in Cuba if we consider it solely as a manifestation of 'international communism' and fail to relate it either to parallel movements in other parts of the underdeveloped world or to the long and tangled story of relations between the United States and Cuba since 1901? If it is to be of any lasting value, the analysis of contemporary events requires 'depth' no less – perhaps, indeed, a good deal more – than any other kind of history; our only hope of discerning the forces actually operative in the world around us is to range them firmly against the past. Unfortunately this is rarely done. When the Korean war broke out in 1950, for example, commentators treated it simply as an episode in the post-war conflict between the communist and the 'free' worlds and the fact that it was part of a far older struggle, reaching back almost exactly a century, for a dominating position in the western Pacific was passed over without so much as a word.[1] It should hardly need saying that a valid assessment must take both aspects into account; but we shall not get far, in the analysis of recent history, unless we realize that those 'aspects of communist rule that form the usual subjects of contemporary writing' are for the most part 'important only as symbols', and that 'deeper historical trends, often forgotten amidst the crises and passions of the day', are usually of more 'lasting significance in explaining the march of men and events'.[2]

In the long run contemporary history can only justify its claim to be a serious intellectual discipline and more than a desultory and superficial review of the contemporary scene, if it sets out to clarify the basic structural changes which have shaped the modern world. These changes are fundamental because they fix the skeleton or framework within which political action takes place.

1. For a brief survey of the Korean question since 1864 see Lee Insang, *La Corée et la politique des puissances* (Geneva, 1959).
2. cf. Ping-chia Kuo, *China. New Age and New Outlook* (2nd ed., Penguin Books, 1960), p. 9.

Examples of them are the changed position of Europe in the world, the emergence of the United States and the Soviet Union as 'super-powers', the breakdown (or transformation) of old imperialisms, British, French, and Dutch, the resurgence of Asia and Africa, the readjustment of relations between white and coloured peoples, the strategic or thermonuclear revolution. About all these subjects there is room for differences of opinion; everyone is free to make his own assessment of their significance. But we are justified in describing them as 'objective' trends, in the sense that, taken together, they give contemporary history a distinctive quality which marks it off from the preceding period. Furthermore, all require study and analysis in depth; they are parts of a process which can never be fully intelligible if it is taken out of its historical context.

In this respect contemporary history is no different in its requirements from other sorts of history. In other respects this is not the case. In particular, the causal or genetic approach, which has become traditional among historians writing under the influence of German historicism, is an unsuitable tool for the contemporary historian who is seeking to define the character of contemporary history and to establish criteria which mark it off from the preceding period. For him the important thing is not to demonstrate (what we all know) that the garment of Clio is a seamless web, but to distinguish the different patterns in which it is woven. A simple example will illustrate what this difference means in practice.

History of the traditional type starts at a given point in the past – the French Revolution, for example, or the Industrial Revolution, or the settlement of 1815 – and works systematically forward, tracing a continuous development along lines running forward from the chosen starting-point. Contemporary history follows – or should follow – an almost contrary procedure. Both methods may take us far back into the past, but it will be a different

past. Thus, in regard to the development of modern industrial society, the contemporary historian will be concerned
less with the gradual extension of industrial processes
from their conventional beginnings with Hargreaves's
spinning-jenny, Arkwright's water-frame, Crompton's
mule, Watts's steam-engine, and Cartwright's powerloom,
than with the substantial differences between the 'first'
and 'second' industrial revolutions; from his point of view
the latter are more significant than the undeniable element
of continuity linking the eighteenth and the twentieth
centuries.[1] In the field of international political history
the differences are no less clear. The historian who starts,
for example, from the situation in 1815 and works forward
step by step and stage by stage will almost inevitably concern himself mainly with Europe, since the problems
which arose directly from the settlement of 1815 were
primarily European problems. For him, therefore, the
main issues will be German and Italian unification, the
so-called 'Eastern Question', the impact of nationalism,
particularly on the Habsburg and Ottoman empires, and
perhaps pan-Slavism – questions which, through their
interactions, culminated (or, it would be more accurate
to say, seemed when looked at from this point of view to
culminate) in the war of 1914 – and events in other parts
of the world will tend to be regarded as peripheral, except
in so far as they can be brought under the heading of
'European expansion'. The historian who takes his stand
not in 1815 but in the present will see the same period in
different proportions. His starting-point will be the global
system of international politics in which we live today and
his main concern will be to explain how it arose. Hence
he will be just as interested in Oregon and the Amur as in
Herzegovina and the Rhine, in the clash of imperialisms
in central Asia and the western Pacific as in the Balkans
or Africa, in the trans-Siberian railway as in the line from
Berlin to Baghdad. Both will survey the same stretch of

1. I shall return to this point later; cf. below, p. 44.

the past, but they will do so with different objects in mind and different standards of judgement.

Although the contemporary historian will necessarily pay attention to different things, it does not follow that his approach need be shallower or his perspective shorter than that of other historians. For a proper understanding of the changeover from a European to a world-wide political system, which is one of the most evident characteristics of the contemporary era, we may, for example, be carried back as far as the Seven Years War, which has been described as 'the first world conflict of modern times'.[1] Or who again, when the Russian occupation of Berlin in 1945 was described as an unprecedented Slavonic advance to the west, paused to recollect that the Russians had already occupied Berlin in 1760? Evidently this is not contemporary history, any more than the campaigns of Suvorov's armies in Italy and Switzerland during the Napoleonic wars are contemporary history; but it is important to be aware of them and to take them into account, if we are to see recent events in perspective. To understand the position of Russia in Asia – which, like the expansion of the United States across the American continent to the Pacific, is one of the preconditions of the modern age – it may be necessary to look back, however briefly, to Yermak's Siberian campaigns in the early 1580s and the astounding advance across Asia which brought Russian explorers and adventurers to the Pacific coast by 1649. And, again, it would be foolish to expect to understand the policy of the United States today, without looking back beyond the 1890s and the Philippine and Cuban wars to the earlier phases of American imperialism which Professor van Alstyne has so brilliantly surveyed.[2]

These few examples are sufficient to show that contemporary history does not signify – as historians have

1. cf. S. F. Bemis, *The Diplomacy of the American Revolution* (2nd ed., Bloomington, 1957), p. 5.

2. cf. R. W. van Alstyne, *The Rising American Empire* (Oxford, 1960).

sometimes contemptuously implied – nothing more than scratching about on the surface of recent events and misinterpreting the recent past in the light of current ideologies. But they also show – which is fundamentally more important – why we cannot say that contemporary history 'begins' in 1945 or 1939, or 1917, or 1898, or at any other specific date we may choose. There is a good deal of evidence, which I shall bring forward later, the cumulative effect of which is to suggest that the years immediately before and after 1890 were an important turning-point; but we shall do well to beware of precise dates. *Contemporary history begins when the problems which are actual in the world today first take visible shape*; it begins with the changes which enable, or rather which compel us to say that we have moved into a new era – the sort of changes, as I have already suggested, which historians emphasize when they draw a dividing line between the Middle Ages and 'modern' history at the turn of the fifteenth and sixteenth centuries. Just as the roots of the changes which took place at the time of the Renaissance may lead back to the Italy of Frederick II, so the roots of the present may lie as far back as the eighteenth century; but that does not make it impossible to distinguish two ages or invalidate the distinction between them. On the other hand, it indicates that there was a long period of transition before the *ethos* of one period was superseded by the *ethos* of the other; and we shall, in fact, find in the following pages that we are involved in large degree in a transitional age in which two periods, the 'contemporary' and the 'modern', uneasily coexisted. It is only now that we seem to be drawing out of this transition into a world whose outline we cannot yet plot.

2

If we associate the concept of contemporary history, as I believe we should, with the onset of a new era, what label should we put on it? The answer is that we shall be well advised at present to avoid a label altogether. It is true that the term 'contemporary history' is provisional and ambiguous, but it is also colourless; and at present, as we begin to emerge from a long period of transition, it is safer to stick to a colourless, if meaningless, appellation rather than to adopt one which is precise but inaccurate. When we can see more clearly the newly emerging constellation of forces it will be time to think of a term which more nearly represents the world in which we live.

It is true that there have already been a number of attempts to find a new formula, but none is entirely satisfactory. They have been made by historians who have perceived, quite correctly, how rickety the conventional threefold division of history into 'ancient', 'medieval' and 'modern' has become. In particular, it has been suggested that, just as the Mediterranean was succeeded by a European age, so now the European has been, or is being, succeeded by an Atlantic age.[1] This scheme, which implies that the central theme of contemporary history is the formation of an Atlantic community, is plausible and attractive; but there are three reasons why we may hesitate before endorsing it. First of all, it is more a political than a historical concept; it took shape as a projection backwards from the Atlantic Charter of 1941 and was not current, so far as I have been able to discover, among historians before the Second World War.[2] Secondly, the sequence 'Mediterranean–European–Atlantic' is as much a reflection of a European point of view as the sequence 'ancient–medieval–modern' which it is intended to replace,

1. cf. O. Halecki, *The Limits and Divisions of European History* (London, 1950), particularly pp. 29, 54, 60 f., 167 f.

2. Among those who gave it currency was the American political commentator, Walter Lippman; from him it passed, in 1945, to the historians, Carlton Hayes, Garrett Mattingly and Hale Bellot, after which it became a fairly widespread concept. For a short account of its lineage, cf. Cushing Strout, *The American Image of the Old World* (New York, 1963), pp. 221 ff.

and for that reason alone it is a dubious appellation for a period one of the most obvious characteristics of which has been a decline in European predominance and a shift of emphasis away from Europe. And, finally, although there is no reason to deny the existence of 'an historic Atlantic economy' of which the countries on the two seaboards of the Atlantic are 'interdependent parts', it is clear beyond all reasonable doubt that the trend in recent times has been for this economic community to get weaker rather than stronger.[1] Careful investigation shows that it was in the period 1785–1825 that the economic bonds between western Europe and America were closest; thereafter they relaxed slowly until 1860, and after 1860 the slackening gathered pace.[2] Today, in spite of the Atlantic alliance, the two seaboards of the Atlantic are economically 'more distant from each other than they were a century ago'; certainly – and from the present point of view significantly – 'the decade of the nineties' was 'the end of one epoch and the beginning of another' in the history of the Atlantic economy.[3]

It would thus seem that there is little justification, for the historian soberly considering the facts, for adopting the view that contemporary history is, in its broader outline, interchangeable with the story of the rise of a new 'Atlantic' era. Indeed, if we base our conclusions on the course of events since 1949, it would be just as easy and just as plausible to argue that the world was moving not into an Atlantic but into a Pacific age. The war in Korea,

1. This is the conclusion of J. Godechot and R. R. Palmer in their brilliant re-examination of the whole question, 'Le problème de l'Atlantique du XVIIIᵉ au XXᵉ siècle', printed in vol. v of the *Relazioni* of the Tenth Congress of Historical Sciences (Florence, 1955), pp. 173–239.

2. ibid., p. 199.

3. This is the conclusion of Brinley Thomas, *Migration and Economic Growth. A Study of Great Britain and the Atlantic Economy* (Cambridge, 1954), pp. 118, 235; cf. also Godechot and Palmer, op. cit., p. 235.

the conflict in Vietnam, problems of Laos – issues which since 1945 have been nearer than anything that has happened in Europe to sparking off a Third World War – the long-drawn-out and unresolved question of Formosa, and the tensions in south-east Asia between Indonesia and Holland and Indonesia and Great Britain, quite apart from the stupendous transformation which has gone on in China since 1949: what, it may be thought, are these but evidence that the axis of world history, which the philosophers of the eighteenth century saw moving from east to west, has taken one further westward leap and completed the circle? But such metahistorical speculations, fascinating as they sometimes are, are better left aside. The simple fact is that we do not have sufficient knowledge to decide such issues. The new period which we call 'contemporary' or 'post-modern' is at its beginning and we cannot yet tell where its axis will ultimately lie. All the labels we put on periods are *ex post facto*; the character of an epoch can only be perceived by those looking back on it from outside. That is why we must be content for the present with a provisional name for the 'post-modern' period in which we live. On the other hand, precisely because we stand outside it and can look back over it from outside, we can see the period which we still call 'modern history' – the European age which Pannikar declared to have begun in 1498 and ended in 1947[1] – as a process with a beginning and an end; and the very fact that we are able to form some notion of the structure and character of this earlier period enables us to establish, by contrast and comparison, some at least of the differentiating features of the period that followed it. It is these features, as I understand them, that will be the subject of the following chapters.

1. cf. K. M. Pannikar, *Asia and Western Dominance* (London, 1953), p. 11.

3

It is true that no sharp line divides the period we call 'contemporary' from the period we call 'modern'. In this we can agree with the upholders of the doctrine of historical continuity. The new world grew to maturity in the shadow of the old. When we first become aware of it, towards the close of the nineteenth century, it is little more than an intermittent stirring in the womb of the old world; after 1918 it acquires a separate identity and an existence of its own; it advances towards maturity with unexpected speed after 1945; but it is only in the very recent past, beginning around 1955, that it has thrown off the old world's tutelage and asserted the inalienable right to decide its own destiny. Its history is therefore a good deal less than the whole history of the period involved – indeed in the early years it is only a very small part of that history – and this is a complicating factor to which we shall return. But if our object is to understand the origins of the age in which we live and the constituent elements which make it so different from the European-centred world of the nineteenth century, we shall hardly be wrong if we say that it is the part that matters most to us.

When we seek to isolate those strands in the history of the period which lead forward to the future, it soon becomes evident – no matter which particular line we choose to follow – that they converge with surprising regularity at the same approximate date. It is in the years immediately preceding and succeeding 1890 that most of the developments distinguishing 'contemporary' from 'modern' history first begin to be visible. No doubt it would be unwise to exaggerate the significance of this – or any other – date as a dividing-line between two periods; it is more like the line of a graph, representing a statistical average with a considerable margin of fluctuation on either side. Even so, it is too well substantiated to be ignored. Before the nineteenth century had closed, new forces were bringing about fundamental changes at practically every level of living and in practically every quarter of the inhabited globe, and it is remarkable, if we examine the literature of the period, how many people were aware of the way things were moving. The ageing Burckhardt in Basel, the English journalist, W. T. Stead, with his vision of the 'Americanization of the world', Americans such as Brooks Adams, even Kipling in the sombre 'Recessional' he wrote for Queen Victoria's jubilee in 1897, are only a few of the more outstanding figures among a multitude who sensed the unsettling impact of new forces: their particular prognostications, the fears and hopes they attached to the changes going on around them, may have proved wrong, but their perception, often dim but sometimes acute, that the world was moving into a new epoch was not simply an illusion.

When we seek to identify the forces which set the new trends in motion, the factors which stand out are the industrial and social revolution in the later years of the nineteenth century and the 'new imperialism' which was so closely associated with it. The nature and impact of these interlocked movements, much debated in recent years, will be examined in the following chapter; here it is sufficient to say that it is only by distinguishing what was new and revolutionary in them − in other words, by emphasizing the differences between the 'first' and the 'second' industrial revolutions and between the 'old' and the 'new' imperialisms − that we can expect to measure the full consequences of their impact. It is also true, of course, that it was some time before these consequences became explicit. None of the changes we shall have to consider in the following pages − neither the transition from a European to a global pattern of international politics, nor the rise of 'mass democracy', nor the challenge to liberal values − was decisive in itself; none alone was sufficient to bring about the shift from one period to another. What was decisive was their interaction. Only when the constellation of political forces, which was still confined to Europe in the days of Bismarck's ascendancy, became involved with other constellations of political forces in other parts of the world; only when the conflict between peoples and governments interlocked with the conflict of classes, which was still not the case in 1914; only when social and ideological movements cut across frontiers in a way (or at least to an extent) that was unknown in the period of national states: only then did it become clear beyond all dispute that a new period in the history of mankind had arrived.

It is from this point of view that the various events which have been picked out as milestones marking the stages in the transition from one epoch to another have to be considered. Among them the war of 1914–18, with the unprecedented dislocations that followed in its train, was the first. For contemporaries and later writers alike, no other event heralded more clearly the ending of an epoch. 'It is not the same world as it was last July,' the American ambassador in London told President Wilson in October 1914;[1] 'nothing is the same.' But though many were to echo his words, it is evident today that they exaggerated the speed of change. In the first place, the end of one epoch is not necessarily coincidental with the onset of another; there may be – and in fact there was – a period of confused and uncertain tendencies in between. In the second place, the recuperative powers of the old European-centred world were formidable. The war of 1914–18 brought into relief the hidden and unresolved tensions which had been gathering strength since the closing years of the nineteenth century; it weakened the framework of society and made it easier for new forces to make themselves felt. But few things are more remarkable than the speed with which after 1919 the threat of radical social upheaval was banished; and it only needed the withdrawal of the United

1. cf. B. J. Hendrick, *The Life and Letters of Walter Hines Page,* vol. III (London, 1925), p. 165.

States into isolation and the elimination of Soviet Russia by revolution and civil war to convince European statesmen that international politics had not, after all, departed substantially from their old pattern. The urge to return to 'normalcy' – an urge which revealed the vitality of the conservative forces stemming from the old world – was one of the most conspicuous features of the decade between 1919 and 1929.

It is obvious to us today that this hankering for a return to pre-1914 conditions and the belief, prevalent between 1925 and 1929, that it had been attained, were illusory. Whatever the appearances to the contrary, the world was in fact moving on. Although by 1925 most economic indexes had reached, if not overtaken, the level of 1913, the war had brought substantial and irreversible changes in the balance of economic power, and in relation to over-all growth the countries which had taken the lead in the pre-war world – Germany, for example, the United Kingdom, France, and Belgium – were falling back.[1] The position in the field of international politics was much the same. Here the shift in balance was masked by the temporary absence of the United States and the Soviet Union, but it never ceased to be the underlying reality and it is difficult today to follow the calculations and manoeuvres of European diplomacy in the inter-war years – from the Little Entente of 1921 to the Non-Intervention Committee of 1936 – without experiencing a feeling of futility only matched, perhaps, by the futility of Athenian politics in the days of Alexander the Great. It was an 'era of illusions'.[2] But the illusions were a potent factor in the history of the period – particularly the illusion that Europe retained the dominant position it had claimed in pre-war days. One result, among many, was that those in charge of

1. cf. W. A. Lewis, *Economic Survey, 1919–1939* (London, 1949), pp. 34–5, 139.
2. The phrase is that of René Albrecht-Carrié, *A Diplomatic History of Europe* (London, 1958), p. 385; cf. also pp. 301–4.

British policy in the thirties were so obsessed by Mussolini and Hitler that they neglected Hirota and Konoye, and when in July 1937 the Japanese began the Second World War which brought the European empires crashing down, they did not realize that the Second World War had begun. Mao Tse-tung was quick to point out the illusion behind this attitude. Japanese policy, he said, was 'directed not only against China', but also against all those countries with interests on the Pacific ocean, and neither England nor the United States would be able to 'remain neutral'.[1] But his words fell on stony ground. The mental horizon of European statesmen – even of those who, like the English, had major interests outside Europe – were still circumscribed by the presuppositions of the old world and dominated by the belief that the only significant things going on as late as 1939 were the things going on in Europe.

No one concerned with the period since 1918 can afford to ignore the persistence of old ways of thought and the conservative resistance to change. In a full-scale history of the period they would loom large. Throughout the years of transition the breakthrough of the new was impeded by the retarding force of the old. At each milestone we can see, if we look back, that the old positions were being eroded and undermined. That is true of the year 1917 which more than one historian has picked out as the decisive

1. Mao's remarks were reported by Edgar Snow, *Red Star over China* (London, 1937), pp. 94, 102. 'We know', Mao continued, 'that not only north China but the lower Yangtze valley and our southern seaports are already included in the Japanese continental programme. Moreover, it is just as clear that the Japanese navy aspires to blockade the China seas and to seize the Philippines, Siam, Indo-China, Malaya and the Dutch East Indies. In the event of war, Japan will try to make them her strategic bases, cutting off Great Britain, France and America from China, and monopolizing the seas of the southern Pacific. These moves are included in Japan's plans of naval strategy, copies of which we have seen.'

turning-point;[1] it is even more clearly true of the slump of 1929. But even after 1945 there were strong 'restorative' tendencies at work and it was only the failure of these that gave the impetus for the decisive leap into a new world. The burying of the age-old Franco-German rivalry, the search for a new statute for western Europe, the recognition of the division between western and eastern Europe which this implied, the outcome of the Suez war of 1956 and Macmillan's 'wind of change' speech early in 1960, were all evidence of a desire to liquidate the old concern before it crashed down in bankruptcy. But more important in the long run was the fact that the issues which were now agitating the world were predominantly new issues, reflecting a situation that had not existed until a few years earlier. By the end of 1960 it may fairly be said that the long period of transition was over.

Even so, we must not think in terms of a clear-cut break. When the decisive changes began towards the close of the nineteenth century, they had done so in a world which, for all its expansiveness and in spite of symptoms of *fin de siècle* malaise, was securely anchored to two fixed points: the sovereign national state and a firmly established social order stabilized by a prosperous property-owning middle class. Both characteristics proved remarkably tenacious. They weathered the storms of two world wars, and are still factors to be reckoned with in the world of today. Concepts such as sovereignty, the national state, and a property-owning democracy, middle-class in structure though expanded by the absorption of large segments of the working class, have been carried over as components of a society essentially different from that of 1914, in much the same way as the Germanic societies of the early European middle ages incorporated elements taken over from Rome. It is possible that these are dying elements,

1. e.g. E. Hölzle, 'Formverwandlung der Geschichte Das Jahr 1917', *Saeculum*, vol. VI (1955), pp. 329–44; H. Rothfels, *Zeitgeschichtliche Betrachtungen* (Göttingen, 1959), p. 11.

mere survivals which will disappear in the course of a few generations, as most of the Roman inheritance eventually became obsolete in Frankish Gaul; it is possible that they will remain – transformed, no doubt, and adapted to new conditions, but powerful and active – as constituent elements of a new society. We do not know and it would be pointless to speculate. All we can say with certainty is that they exist as counterbalancing factors in the contemporary situation, as elements of continuity which offset the elements of discontinuity and change. They indicate – what any historian with experience of similar changes in the past would expect – that the world which has emerged is neither sharply cut off from the world out of which it emerged nor simply a continuation of it; it is a new world with roots in the old.

4

If the retarding influence of conservative forces fighting to preserve as much as possible of the old European-centred world was one factor affecting the process of transition, another factor was the disruption of the heart of Europe through the rivalries and conflicts of the European powers between 1914 and 1945. No aspect of recent history has been more fully discussed. For most European historians the disputes and rivalries that gathered momentum after 1905 marked the beginning of the great civil war in which Europe, caught in the toils of its own past, encompassed its own destruction, and it was the failure of Europe to solve its own problems – in particular, the long-standing problems of nationalism – that ushered in a new age.

No one would deny that this view of contemporary history, with its emphasis on Europe and on the continuity of developments within Europe, illuminates certain aspects of the history of the period. The real question is whether it is adequate as a key to the process of transition as a whole. The years between 1890 and 1960 confront us with two interlocking processes, the end of one epoch and the beginning of another, and the conflicts of the European powers undoubtedly played a large part in the former. What we have to ask is whether historians who have made Europe the pivot of their story have not concentrated too exclusively on the old world that was dying and paid too little attention to the new world coming to life. It is no doubt true that, but for the wars which brought the old world crashing down, the birth of the new world would have been more protracted and difficult. Their course and outcome also throw light on the post-war situation in Europe. But as soon as we extend our view from Europe to Asia and Africa, the position is different. There, as we shall see,[1] the conflicts and rivalries of the European powers were a contributory factor; but they do not help us to understand the character of the new world which emerged after 1945, any more than they explain the origins and growth of the forces that shaped it during the preceding fifty years. An interpretation which concentrates on the European predicament, in short, is too narrow for a process which was world-wide; it may not be wrong within its own limits, but it is misleading in balance and perspective.

Nor shall we understand the course of events in Europe itself, if we dissociate it from the world-wide process of change which began around 1890. The European conflicts of the first half of the twentieth century were more than a continuation of earlier European conflicts. From the end of the nineteenth century Europe was involved simultaneously in the problems inherited from its own past and in a process of adaptation to a new world situation, and both aspects of its history must be taken into account. For this reason it is easy to place disproportionate emphasis on the unsolved problems of nationalism, as they had developed in Europe since 1815. These problems, particularly the growth of German nationalism, were one factor in the

1. cf. below, pp. 154 ff.

situation; but equally important was the awareness – prominent in the minds of writers such as Hans Delbrück, Rudolf Kjellén, Paul Rohrbach, and Friedrich Naumann – that the position of Europe in the world was changing and that it would be irretrievably lost unless something were done to restore it. We can see this conviction emerging and gathering strength – particularly but not exclusively in Germany – during the 1890s, as a response to the new imperialism of the period, and we can see also how it was caught up and identified with the realization of German national aims. But it was never simply an expression of German nationalism. Rather its foundation was the conviction that policies which aimed merely to defend established positions were fighting a losing battle, and that a more positive reaction was necessary. This reaction has been called 'the last attempt to reorganize modern Europe'.[1] The form it took was an attempt to weld together in the heart of Europe the core of a German-dominated empire strong enough to compete on terms of equality with the other great world powers of the time, imperial Russia, the United States, and the British empire. Its outcome was the wars of 1914 and 1939.

We shall have more to say later of the way this German attempt to reshape Europe affected the transition from a European to a world-wide system of international politics.[2] Here it concerns us in so far as it throws light on the origins of those forces which were later to take shape as Fascism and National Socialism. These forces were a characteristic by-product of the old world in decline. In 1914 they were still far weaker than the forces stemming from the past, particularly the force of European nationalism. But the further disintegration proceeded, the more they gathered strength. Divided at first among a number of small eccentric splinter-groups at loggerheads with bourgeois society – the so-called 'revolutionaries of the right' or 'right-wing radicals', of whom Moeller van den

1. cf. Halecki, op. cit., pp. 167, 182.    2. cf. below, pp. 111–16.

Bruck is perhaps the typical example[1] – they drew strength from the turmoil and distress in Europe after 1918, until finally, with the onset of depression in 1929 and the sharpening antagonism between capitalism and communism, they became a major political force. Resistance to Hitler from within Europe was incomparably weaker in 1939 than resistance to Germany had been in 1914. The reason was that the national spirit which had sustained Europe from 1914 to 1918 had lost its *élan* and Fascist ideas had won a following in most European countries. Their emergence clouded and complicated the central issues of the age. Henceforward we find ranged against the conservative forces fighting tenaciously to maintain the old European world not only those on the left intent on replacing it by a new society but also those on the extreme right whose object was to reshape Europe in a form better able to withstand the onslaught of revolutionary conditions; and between these poles there was room for an infinite variety of groupings and regroupings.

The temptation to treat the ensuing complications as the substance of contemporary history is very great. To do so would be to fail to see the wood for the trees. The impact of Fascism in its various forms multiplied the possibilities of tactical manoeuvre, but it is not clear how substantially it affected the transition from one epoch of history to another. So far as the world situation was concerned, the consequences of National Socialism and Fascism may be brought under three headings, all of them indirect. First, they divided the forces fighting to defend the old order and so weakened and eventually disrupted the delaying action which had been so effective a brake on radical change for ten years before 1929.

1. The early chapters of O. E. Schüddekopf, *Linke Leute von Rechts* (Stuttgart, 1960) contain an informative account of the 'rebellion of the youth of Europe against tradition, convention and a petrified order', and more specifically of the origins of right-wing radicalism in Germany; for Moeller van den Bruck, cf. pp. 35–7.

Secondly, they emerged during the 1930s as the most formidable challenge to the *status quo* – far more immediately dangerous than left-wing radicalism or colonial disaffection – with the result that they drove the other two alignments, the conservative right and the socialist (and communist) left, into a temporary alliance which was one main reason for the enhanced power exercised by the latter after 1945. And finally, by deflecting attention from other issues and focusing it on the 'Fascist threat' in Europe, they helped to accelerate change in other parts of the world. Thus the long series of concessions in the Far East resulting from British preoccupation with Mussolini in the Mediterranean and Hitler in Europe encouraged and facilitated the policies of Japan, which were to prove one of the most powerful solvents of the old order in Asia.

In all these ways Fascism and National Socialism, which claimed to be the only effective instruments for shoring up the old world – and which won mass support on that score – turned out, by a peculiar irony of history, to be instruments in its collapse. They played a part in the process of transition as factors forcing forward the march of events; but their positive contribution to the new world arising amidst the ruin of the old was small. It would be a superficial analogy, for example, to derive the 'guided democracy' of Indonesia or the 'basic democracy' of Pakistan from the Fascist corporate state or to attribute the political structure of Argentina after 1945 to Perón's visit to Italy between 1939 and 1941 rather than to see it in the context of the social changes in Latin America inaugurated by the Mexican revolution of 1910. If we wish to understand why, among the many possibilities opened up by the collapse of Germany and Japan in 1945, certain ones materialized and others did not – why, for example, the fall of Japan did not result in the restoration of the British, French and Dutch empires in eastern Asia – we must turn to developments which historians have too easily banished to the outer margins of history and which are only now slowly finding their way back to the centre. Today it is evident that much we have been taught to regard as central is really peripheral and much that is usually brushed aside as peripheral had in it the seeds of the future. Looked at from the vantage-point of Dien Bien Phu, for example, Amritsar stands out with new and unaccustomed prominence among the events of 1919.

It is no doubt true that, down to 1945, the end of the old world was the most conspicuous aspect of recent history; it engrossed the attention of contemporaries and blinded them to the importance of other aspects. But it is the business of the historian, looking back over events from a distance, to take a wider view than contemporaries, to correct their perspectives, and to draw attention to developments whose long-term bearing they could not be expected to see. For the most part they have made little use of their opportunity; indeed it sometimes seems as though they are in danger of being frozen for ever in the patterns of thought of the years 1933–45. In part, no doubt, this is due to the fact that many historians are still emotionally involved in the death-agonies of the old world, which they feel more deeply than the birth-pangs of the new; it is due, also, to the fact that, until very recently, we were unable to stand outside the period of transition and look back over it as a whole. Today that is no longer the case. If, as I have tried to indicate, the long transition from one age to another is now over, if we can say that between 1955 and 1960 the world moved into a new historical period, with different dimensions and problems of its own, it should no longer be impossible to restore the balance between the old world which has passed and the new world which has emerged.

To do so is also a matter of urgent practical necessity. It would be dangerously misleading to assume that the phenomena of transition, which were the mark of the period 1918–56, will be characteristic of the new era. The rising generation will inevitably look back over the twentieth century with different priorities from ours. Born into a world in which – as all present indications suggest – the major questions will not be European questions but the relationships between Europe, including Russia, and America and the peoples of Asia and Africa, they will find little relevance in many of the topics which engrossed the attention of the last generation. The study of contemporary history requires new perspectives and a new scale of values. We shall find more clues, for example, in Nkrumah's autobiography than in Eden's memoirs, more points of contact in the world of Mao and Nehru than in that of Coolidge and Baldwin; and it is important to remember that, while Mussolini and Hitler were prancing and posturing at the centre of the European stage, changes were going on in the wider world which contributed more fundamentally than they did to the shape of things to come. The tendency of historians to dwell on those aspects of the history of the period which have their roots in the old world sometimes seems to hamper rather than to further our understanding of the forces of change. Here we shall try to strike a different balance. We shall not forget that the end of one epoch and the birth of another were events happening simultaneously within the same contracting world; but it is with the new epoch growing to maturity in the shadow of the old that we shall be primarily concerned.

5

Every day that passes brings new indications that the long period of transition with which this book is primarily concerned has ended, and that the events of the very recent past belong to a new and unsurveyable phase of history. For that reason no attempt will be made to deal with them here, still less to forecast the shape of things to come. That does not mean that I am unaware that developments in many areas of the world have moved beyond the point –

roughly the end of the fifties – I have taken as a terminus; it means only that as yet they are hardly ready for historical appraisal. The sort of writing which attempts to wring the last ounce of meaning out of developments such as the ideological conflict between China and the U.S.S.R. or the political instability of newly emancipated Africa oversteps the limits of historical analysis; the range of possibilities is still so great that any attempt to discuss them is bound to be hypothetical and speculative.

If we wish to mark the opening of this new period – which is, of course, the period of 'contemporary' history in the strict sense of the word – the end of 1960 or the beginning of 1961 is as good a date as any and it is tempting to take the start of the Kennedy administration in the United States as a convenient point for registering the break. This was the first occasion on which decision-making at the highest level passed into the hands of a generation which had not been involved in politics before 1939 and which was not conditioned – in the way, for example, that Sir Anthony Eden's reactions had been conditioned in 1956 – by 'pre-war' attitudes and experience. Nevertheless it would be a mistake to pay too much attention to the personal factor. It was rather a question of cumulative trends which came to a head around the time of Kennedy's accession to power, and so far as his administration registered a change, it would be nearer the truth to regard it as a reflection rather than a cause of a new situation. By the end of 1960 changes which had been taking shape since the death of Stalin in 1953, had reached the stage of crystallization. At the same time, in every quarter of the globe, new problems had emerged which had little direct connexion with the problems of the period of transition.

Already by 1958, 'a turning-point in modern Asian history',[1] it was evident that international politics were operating in a new context. The 'cold war', which had

1. cf. M. Brecher, *The New States of Asia* (London, 1963), p. 73.

claimed first place from 1947 to the Suez war and the Hungarian revolt of 1956, ceased to be the dominant issue, and in the post-Suez and post-Hungary atmosphere of stalemate, a decisive shift of focus took place. Among other things, 'new local points of friction were coming into existence' which in the long run 'could not fail to influence wider alignments'.[1] In the communist *bloc*, the ideological controversy between China and the Soviet Union, simmering since 1957, came to a head in 1959.[2] In Asia the common front established at Bandung in 1955 gave way to territorial disputes between China, on the one hand, and India, Burma and Pakistan, on the other. In Africa, where 1958 was also 'a year of growing tension',[3] the dismantling of European colonialism had hardly been completed before the economic and political problems of independence made themselves felt. In western Europe the Rome treaties of 1957 registered the conclusion of the first stage in the move towards new forms of regional integration. What was common to all these issues was that they marked the emergence of a new phase of history. At mid-century the world was still grappling with the problems of transition; ten years later it was settling down to a new pattern.

To discuss this new pattern in detail would require another and in many ways a very different book. Nevertheless it is not difficult to pick out some of the more obvious ways in which it differed from the old. The most conspicuous was the new prominence of China, unmistakably advancing towards the status of a world-power. More fundamental was the change in relations between the communist and non-communist worlds, a

1. For an admirable analysis of the new situation, cf. R. F. Wall in *Survey of International Affairs, 1956–1958* (London, 1962), pp. 400 f.

2. D. S. Zagoria, *The Sino-Soviet Conflict, 1956 – 1961* (Princeton, 1962).

3. cf. C. E. Carrington in *Survey of International Affairs, 1956–1958*, p. 444.

change due not to settlement of outstanding problems, but to the realization that the old issues were no longer the insistent issues, and that in any case there was no practical alternative, in the world as it was, to some form of co-existence. The result was an abatement of ideological strife and a growing impatience with ideologies which originated in the European past and were no longer congruent with the realities of a world which had ceased to be centred on Europe. The counterpart to this emancipation from the tyranny of outworn concepts was the appearance of 'neutralism' as a new political principle. The sudden, unexpected emergence of new problems in the aftermath of Asian and African emancipation – above all, the problems caused by the growing disparity between the industrialized and the underdeveloped countries – tended to cut across old alignments and to produce new divisions without parallel in the old world. And although, on the surface, the most striking feature of the new situation was the pullulation of new nationalisms, more significant of the rise of new patterns was the evidence of awareness that technological progress required larger groupings, and that the traditional national unit, which was another legacy of nineteenth-century Europe, was an inadequate basis for coping with the problems of technological society. The tendency to form new regional groupings was world-wide; it was at work not only in eastern and western Europe, where it was seen in the establishment of 'Comecon' and the western European common market, but also in Latin America, in the Arab world, and in Africa, where many of the newly emerging states 'adopted the federal idea even before full independence'.[1] Finally, there was a general realization that, so long as the existing thermonuclear balance of power continued to exist, the new patterns could not be altered in any substantial way by

1. cf. P. Calvocoressi, *World Order and New States* (London, 1962), p. 100.

recourse to war. Thus a world of great regional blocks seemed to be arising, different in almost all its preconditions from the world of nation-states of thirty or forty years earlier – a world in which communism and capitalism would figure more as alternative systems than as conflicting ideologies, and in which the great overriding issues, from which no one could contract out, would be the problems of poverty, backwardness, and overpopulation.

It is none of our business to try to depict the lines of development of this new world or the probable impact of other more fundamental changes. There is every likelihood that atomic energy, electronics, and automation will affect our lives even more fundamentally than the industrial revolution and the scientific changes at the close of the nineteenth century. As yet, however, we cannot hope to measure their impact and it would be unprofitable to attempt to do so. But it is only necessary to compare the world situation at mid-century and the world situation today to realize that we have crossed the threshold of a new age. In 1949, for example, the expansion of communism into China and eastern Europe could still be thought of as a temporary, reversible advance; when Dulles died ten years later it was clear that it was there to stay, and the hope of forcing it back, which was a dominant theme of the period from 1947 to 1958, had given way to speculation on the possibility of evolution within the communist world as the basis for a *modus vivendi*.

Such changes were more than superficial. They marked the starting-point of new lines of development leading into a new era. When communism, which down to 1939 had been confined as a political system to one country and to about eight per cent of the world's inhabitants, became the political system of almost one-third of the population of the globe, and when capitalism, which between the wars had directly or indirectly controlled nine-tenths of the world's surface, was reduced by the rise of the neutralist *bloc* to a minority position in the world as a whole and in the United Nations – which was the case by 1960[1] – the old political framework was irretrievably shattered. It was not that the new ideas triumphed – for the most part they did not – but rather that the sheer attrition of events made it necessary to come to terms with new circumstances. Even then, of course, there was a residue of problems left over from the old world. But the balance had changed and the order of priorities was no longer the same. Nothing is more noticeable around 1958 than the liquidation of what, up to that time, had been regarded as the essential problems of the twentieth century. By comparison with the insistent problems of over-population and under-development in Asia and Africa, issues such as German unification fell into the background, and the permanence of the Oder-Neisse frontier was tacitly accepted. In this respect, as in many others, the new world seemed to be moving in directions almost the contrary of the old. The problems anchored in the European past were losing urgency, the values of the age of European nationalism were crumbling, and the focus of interest had passed from the Atlantic, where the North Atlantic Treaty Organization had become an almost meaningless survival, to the Pacific. In 1950 Asia and Africa had been continents at the end of colonialism; a decade later they had passed into the post-colonial age, and with the end of colonialism a new phase of world history had begun.

Whether this new phase represents an advance is not, of course, the relevant point. Many of the expectations bound up with the ending of colonialism were extravagant and unlikely to be fulfilled; and the long series of *coups*, beginning in Burma and Pakistan in 1958 and continuing in quick succession to the upheavals in Nigeria and Ghana

1. At the end of 1960 Adlai Stevenson admitted that, 'due to the admission of so many new countries, the United States and the western democracies no longer control the United Nations'; cf. R. B. Stebbins, *The United States in World Affairs, 1960* (New York, 1961), p. 357.

in 1966, only registered the intractability of the problems facing the ex-colonial peoples. The essential feature of the new age was that the world was integrated in a way it had never been before; and this meant that no people, however small and remote, could 'contract out'. A century ago the Taiping rebellion in China was a distant event, which left Englishmen and Europeans untouched; today what happens in Laos or Vietnam is as likely to spark off the Third (and last) World War as Balkan affairs were to initiate the chain of events leading to the First World War in 1914.

The new period, at the beginning of which we stand, is the product of basic changes in the structure of national and international society and in the balance of world forces. It is a period of readjustment on a continental scale, and its emblem is the mushroom cloud high above Hiroshima and Nagasaki, the nuclear pile in which the old certitudes were consumed for ever. It is also a period which has experienced a breakthrough in scientific knowledge and achievement, and an alliance between science and technology, which has the power to change for all time the material basis of our lives on a scale inconceivable only fifty years ago, but which at the same time has brought us face to face with the possibility of self-extinction. It is, in short, a period of explosive new dimensions, in which we have been carried with breathtaking speed to the frontiers of human existence and deposited in a world with unparalleled potentialities but also with sinister undercurrents of violence, irrationality, and inhumanity. The views we take of this new world may differ, and it is easy to speculate on the course of development it will follow; all we can safely say – with Valéry[1] – is that, if historical experience is anything to go by, the outcome will betray all expectations and falsify all predictions.

1. cf. Paul Valéry, *Collected Works*, vol. x (London, 1962), pp. 71, 113, 116, 126–7.

## II

# THE IMPACT OF TECHNICAL AND SCIENTIFIC ADVANCE

*Industrialism and Imperialism as the Catalysts of a New World*

WHEN we seek to pinpoint the structural changes which lie at the roots of contemporary society, we are carried back to the last decade of the nineteenth century; and there we come to a halt. Even the most resolute upholder of the theory of historical continuity cannot fail to be struck by the extent of the differences between the world in 1870 and the world in 1900. In England, where the industrial revolution had begun early and advanced in a steady progression, the fundamental nature of the changes after 1870 is less apparent than elsewhere; but once we extend our vision to cover the whole world, their revolutionary character is beyond dispute. Even in continental Europe, with perhaps the sole exception of Belgium, industrialization was a product of the last quarter rather than of the first two-thirds of the nineteenth century; it was a consequence, rather than a concomitant, of the 'railway age', which by 1870 had provided the continent with a new system of communications. Across the Atlantic the civil war had proved a major stimulus to industrialization; but it was after the ending of the civil war in 1865 and the uneasy post-war interlude spanned by the presidencies of General Grant (1868–76) that the great industrial expansion began which transformed beyond recognition the society de Tocqueville had known and described. When in 1869 the first railroad to span the American continent was completed at a remote spot in Utah, the United States 'ceased to be an Atlantic country

43

in order to become a continental nation' of a new, highly industrialized pattern.[1]

What happened in the closing decades of the nineteenth century was not, however, simply an expansion of the process of industrialization which had begun in England a century earlier, until it became world-wide. I have already referred to the distinction between the first industrial revolution and the second, or (as it is sometimes called) between the 'industrial' and the 'scientific' revolutions. It is, of course, a clumsy distinction, which does less than justice to the intricacy of the historical facts; but it is a real one. The industrial revolution in the narrower sense – the revolution of coal and iron – implied the gradual extension of the use of machines, the employment of men, women, and children in factories, a fairly steady change from a population mainly of agricultural workers to a population mainly engaged in making things in factories and distributing them when they were made. It was a change that 'crept on', as it were, 'unawares',[2] and its immediate impact, as Sir John Clapham made clear, can easily be exaggerated. The second industrial revolution was different. For one thing, it was far more deeply scientific, far less dependent on the 'inventions' of 'practical' men with little if any basic scientific training. It was concerned not so much to improve and increase the existing as to introduce new commodities. It was also far quicker in its impact, far more prodigious in its results, far more revolutionary in its effects on people's lives and outlook. And finally, though coal and iron were still the foundation, it could no longer be called the revolution of coal and iron. The age of coal and iron was succeeded, after 1870, by the age of steel and electricity, of oil and chemicals.

1. cf. J. Godechot and R. R. Palmer, 'Le problème de l'Atlanti-que', *Tenth International Congress of Historical Sciences, Relazioni*, vol. v (Florence, 1955), p. 186.

2. C. P. Snow, *The Two Cultures and the Scientific Revolution* (Cambridge, 1959), p. 27.

1

The technical aspects of this revolution do not concern us here, except in so far as is necessary in order to understand its effects outside the spheres of industry, science, and technology. It would nevertheless be difficult to deny that the primary differentiating factor, marking off the new age from the old, was the impact of scientific and technological advance on society, both national and international. Even on the lowest level of practical everyday living it is surely significant that so many of the commonplace objects which we regard as normal concomitants of civilized existence today – the internal combustion engine, the telephone, the microphone, the gramophone, wireless telegraphy, the electric lamp, mechanized public transport, pneumatic tyres, the bicycle, the typewriter, cheap mass-circulation newsprint, the first of the synthetic fibres, artificial silk, and the first of the synthetic plastics, Bakelite – all made their appearance in this period, and many of them in the fifteen years between 1867 and 1881; and although it was only after 1914, in response to military requirements, that intensive aircraft development began, the possibility of adapting the petrol-driven internal combustion engine to the aeroplane was successfully demonstrated by the brothers Wright in 1903. Here, as elsewhere, there was necessarily a time-lag before the problems of large-scale production were solved, and some of the things we have come to regard as normal – radio and television among them – obviously belong to a later phase.[1] Nevertheless, it can fairly be said that, on the purely practical level of daily life, a person living today who was suddenly

1. In the same way, of course, atomic physics, the industrial use of atomic particles, and the exploitation of atomic energy, both for warlike and for peaceful purposes, are twentieth-century developments; but even here the theoretical foundations were laid by the discoveries of Becquerel, Madame Curie, and J. J. Thomson at the close of the nineteenth century.

put back into the world of 1900 would find himself on familiar ground, whereas if he returned to 1870, even in industrialized Britain, the differences would probably be more striking than the similarities. In short, it was around 1900 that industrialization began to exert its influence on the living conditions of the masses in the west to such an extent that it is hardly possible today to realize the degree to which even the well-to-do in the previous generation had been compelled to make shift.

The basic reason for this difference is that few of the practical inventions listed above were the consequence of a steady piecemeal development or improvement of existing processes; the overwhelming majority resulted from new materials, new sources of power, and above all else from the application of scientific knowledge to industry. Down to 1850, for example, steel 'was almost a semi-precious material' with a world production of eighty thousand tons, of which Great Britain made half. The discoveries of Bessemer, of Siemens, and of Gilchrist and Thomas, completely transformed the situation, and by 1900 production had reached 28 million tons. At the same time the quality, or rather the toughness, of the metal was vastly improved by the addition of nickel – a result only possible as a consequence of a process of extracting nickel discovered by Ludwig Mond in 1890. Thus, for all practical purposes, nickel may be accounted a new addition to the range of industrial metals, though it had, of course, been in small demand before. The same applies even more directly to aluminium, which had hitherto been too expensive to be put to common use. With the introduction of the electrolytic process, developed in 1886, its production became a commercial proposition and a new constructional material which was soon to be of first-rate importance – for example, in the nascent aircraft industry – became readily available for the first time.

These advances, and others of a similar character, which were themselves the foundation for further progress, were the result of more fundamental changes still: namely, the introduction of electricity as a new source of light, heat, and power, and the transformation of the chemical industry. Electrolysis, so important in the extraction of copper and aluminium and in the bulk production of caustic soda, only became a practical proposition when electric power became generally available; and the same was true of other electrochemical developments. The electrical and chemical industries of the late nineteenth century were therefore not only the first industries to originate specifically in scientific discovery, but in addition they had an unprecedented impact, both in the speed with which their effects were felt and in the range of other industries they affected. A third new industry with the same revolutionary qualities was petroleum. Here was a source of power equivalent to coal and electricity, and later the raw material of the vast and extending range of petrochemicals. From this point of view the foundation of Rockefeller's Standard Oil Company in 1870 may be regarded in many ways as the symbol of the opening of a new age. By 1897, according to the celebrated American character, Mr Dooley, Standard Oil had a branch in every hamlet in America from the Atlantic to the Pacific coasts, and by this date – although the internal combustion engine was still in its infancy – the United States was already exporting oil to the annual value of $60 million.[1] The impact of electricity was even more spectacular, its stages being marked by Siemen's invention of the dynamo in 1867, Edison's invention of the incandescent bulb in 1879, the opening of the world's first electric power plant in New York in 1882, the establishment of A.E.G. in Germany in 1883, and the construction of the first hydroelectric plant in Colorado in 1890. Even as late as 1850 no one would have foretold the exploitation of electricity as a

1. At this time, however, United States production still lagged behind that of Russia, which, with an annual production of some six million tons, accounted for half of the total world output.

large-scale source of power; but when it passed into common usage, the face of the world was changed. 'Communism', Lenin was shortly to say, 'equals Soviet power plus electrification.'[1]

Another field in which the progress achieved during this period was to be of inestimable future importance was medicine, hygiene, and nutrition. In these branches of knowledge it is perhaps true that the closing decades of the nineteenth century were a less closely defined epoch; but if in some cases the basic experimentation had been made earlier, it was largely after 1870 that its general application took place. Because of prejudice and resistance where the human body was concerned, chloroform only slowly came into use after the middle of the nineteenth century, although it had been discovered as far back as 1831; and in the same way, though carbolic acid was discovered in 1834, the use of antiseptics only became general after Lister began to employ them in Glasgow in 1865. But the main reason why medicine in the mid-nineteenth century was still largely pre-scientific was the fact that the modernization of pharmacy had to await the completion of more fundamental advances in chemistry, and the position in other closely related branches of knowledge was similar. The great age of bacteriology after 1870, associated with the names of Pasteur and Koch, owed its impetus to the development of the new aniline dyes, which made possible the identification of a vast range of bacteria by differential staining methods. Microbiology, biochemistry, and bacteriology all now emerged as new sciences, and among their more significant results were the production of the first of the antibiotics, Salvarsan, in 1909, the discovery of vitamins and of hormones in 1902, and the identification of the mosquito as the carrier of malaria by Sir Ronald Ross in 1897. Aspirin was first marketed in 1899. At the same time anaesthesia, in conjunction with the general use of

1. C. Hill, *Lenin and the Russian Revolution* (London, 1947), p. 199.

antiseptic and aseptic techniques, was revolutionizing medical practice.

The new chemical and physiological knowledge also brought about a revolution in agriculture which was vitally necessary as a counterpart to the upward sweep of the human demographic curve that followed the advance in medicine. The bulk production of basic slag as an artificial fertilizer became possible as a by-product of the new steel-making processes. New methods of food preservation, based on the principles of sterilization and pasteurization used in medical practice, made possible the bulk conservation of foodstuffs and the provision of cheap and stable supplies to the growing world population. As a result of Pasteur's researches the pasteurization of milk for general consumption became usual from about 1890.

It would be hard to exaggerate the importance of these improvements at a time when industrial developments were changing the structure of society and the whole pattern of everyday life. The food-canning industry, helped by new processes of tin-plating, now got into its stride, and the sale of canned vegetables rose from four hundred thousand cases in 1870 to fifty-five million in 1914. Other factors which facilitated the provision of cheap foods for the growing industrial populations were the completion of the main railroad systems, the development of steamships of large tonnage, and the perfection of the techniques of refrigeration. In Europe the piercing of the Alps by the Mont Cenis and St Gotthard tunnels in 1871 and 1882 reduced the journey from Italy and the Mediterranean to France and Germany from days to hours and permitted the bulk import into the industrialized north of southern and subtropical fruits and vegetables. In Canada the completion in 1885 of the Canadian Pacific Railway opened up the great prairies. Refrigerator wagons were in use by 1876, rushing chilled meat from Kansas City to New York, and refrigerated ships carried it to Europe. Consignments of Argentine beef became available in

Europe in good condition from 1877; the first shipload of frozen New Zealand mutton arrived on the English market in 1882. From 1874 the United States provided more than half the total British wheat consumption. Meanwhile, the opening of the Suez Canal in 1869 had cut down the distance between Europe and the Orient, and the traffic it carried multiplied threefold between 1876 and 1890. Colonial and overseas products, such as tea from India and coffee from Brazil, appeared in bulk on the European markets, and Argentina became a main exporter of meat. The combined result was to set in train something not far short of a revolution in the methods of feeding an industrialized and urbanized population.

2

The scientific, technological, and industrial changes I have briefly recapitulated are the starting-point for the study of contemporary history. They acted both as a solvent of the old order and as a catalyst of the new. They created urban and industrial society as we know it today; they were also the instruments by which industrial society, which at the close of the nineteenth century was still for all practical purposes confined to western Europe and the United States, subsequently expanded into the industrially undeveloped parts of the world. Technology, it has been observed,[1] is the branch of human experience people can learn most easily and with predictable results.

The new industrial techniques, unlike the old, necessitated the creation of large-scale undertakings and the concentration of the population in vast urban agglomerations. In the steel industry, for example, the introduction of the blast furnace meant that the small individual enterprise employing ten or a dozen workmen quickly became an anachronism. Furthermore, the process of industrial consolidation was accentuated by the crisis of over-produc-

1. Snow, op. cit., p. 42.

tion which was the sequel of the new techniques and the immediate cause of the 'great depression' between 1873 and 1895. The small-scale family businesses, which were typical of the first phase of industrialism, were in many cases too narrowly based to withstand the depression; nor had they always the means to finance the installation of new, more complicated and more expensive machinery. Hence the crisis, by favouring rationalization and unified management, was a spur to the large-scale concern and to the formation of trusts and cartels; and the process of concentration, once begun, was irreversible.

It went ahead most rapidly in the new industries, such as chemicals, but soon spread in all directions. In England at this time Brunner and Mond were laying the foundations of the vast I C I combine. In Germany the great Krupps steel undertaking, which had employed only one hundred and twenty-two men in 1846, had sixteen thousand on its pay-roll in 1873 and by 1913 was employing a total of almost seventy thousand. Its counterpart in France was Schneider-Creusot, employing ten thousand in 1869; its counterpart in Great Britain was Vickers-Armstrong. In the United States Andrew Carnegie was producing more steel than the whole of England put together when he sold out in 1901 to J. P. Morgan's colossal organization, the United States Steel Corporation. But these were the giants, and in many respects the average performance, as illustrated by the German statistics, is more informative.[1] Here, in the period between 1880 and 1914, the number of small industrial plants, employing five workmen or less, declined by half, while the larger factories, employing fifty or more, doubled; in other words, the number of industrial units declined, but those that remained were substantially larger and employed no less than four times the total of industrial workers recorded for 1880. Furthermore, outworkers,

1. For the following cf. J. H. Clapham, *The Economic Development of France and Germany, 1815–1914* (Cambridge, 1936), pp. 287–8, 290–1, 294, 297.

domestic weavers and the like, who had still been a considerable element in the German textile industries in the early days of the Second Empire – in 1875 nearly two-thirds of the cotton weavers in Germany were domestic outworkers – were virtually eliminated by 1907, as industrial concentration gathered pace. In short, the workers were being gathered into factories and the factories concentrated in industrial towns and urban areas.

The process of herding mill-hands and factory workers into fewer but larger combines was common to all the industrialized countries. It completely changed their physiognomy. The towns devoured the villages and large cities grew faster than small ones. Areas like the Ruhr valley in Germany and the 'Black Country' of the English midlands became sprawling belts of contiguous urban development, divided theoretically by artificial municipal boundaries but otherwise without visible break. A further factor hastening and accentuating the influx into the cities was the agricultural crisis caused by the large-scale import of cheap foodstuffs from overseas. The result was the proliferation of social conditions unknown at any time in the past, the rise of what has usually been called a 'mass society'. As a consequence of the progress of hygiene and medicine the death rate, which had been virtually static between 1840 and 1870, declined abruptly in the following thirty years in the more advanced countries of western Europe – in England, for example, by almost one-third from twenty-two to a little over fifteen per thousand – and the population soared. Compared with an increase of thirty millions between 1850 and 1870, the population of Europe – taking no account of emigration, which drew off forty per cent of the natural increase – rose by no less than one hundred million between 1870 and 1900.

It is a striking confirmation of the shift that was taking place that the whole of this immense increase in population was absorbed by the towns. In Germany, where the census of 1871 recorded only eight cities of over one hundred thousand inhabitants, there were thirty-three by the end of the century and forty-eight by 1910. In European Russia the number of towns in this category had risen by 1900 from six to seventeen. By this time also one-tenth of the inhabitants of England and Wales had been drawn into the vortex of London, and in the United States – although three million square miles of land were available for settlement – nearly half the population was concentrated on one per cent of the available territory and one-eighth lived in the ten largest cities. Whereas before the revolution of 1848 Paris and London were the only towns with a population exceeding one million, the great metropolis now became the hub of industrial society. Berlin, Vienna, St Petersburg, and Moscow in Europe, New York, Chicago, and Philadelphia in the United States, Buenos Aires and Rio de Janeiro in South America, and Tokyo, Calcutta, and Osaka in Asia, all topped the million mark, and it is significant that the emergence of great metropolitan centres was world-wide and that in this respect at least Europe no longer stood out as exceptional.[1]

This, without doubt, was the second most conspicuous aspect of the revolution that was taking place. If its first consequence was to change for all time the social structure of industrial society, its second was to achieve with fantastic speed the integration of the world. This was noted, as early as 1903, by the German historian, Erich Marcks. 'The world', wrote Marcks, 'is harder, more warlike, more exclusive; it is also, more than ever before, one great unit in which everything interacts and affects everything else, but in which also everything collides and clashes.'[2]

1. These developments are surveyed in my contribution to the *Propyläen Weltgeschichte*, vol. VIII (Berlin, 1960), p. 709.

2. E. Marcks, *Die imperialistische Idee in der Gegenwart* (Dresden, 1903). This lecture was reprinted under the title 'Die imperialistische Idee zu Beginn des 20 Jahrhunderts' in the second volume of Marcks's essays, *Männer und Zeiten* (Leipzig, 1911); cf. ibid., p. 271.

This does not imply, of course, that Europe had lost, or was losing, its pre-eminence; on the contrary, the rapidity and extent of their industrialization increased the lead of the European powers and enhanced their strength and self-confidence, and with the sole, if weighty, exception of the United States, the gap between them and the rest of the world widened; even the so-called 'white' dominions, Canada, Australia, and New Zealand, lagged far behind in 1900, and the industrialization of Japan, however remarkable in its own context, remained small by European standards until after 1914. But it is also true that the voracious appetite of the new industrialism, unable of its very nature to draw sufficient sustenance from local resources, rapidly swallowed up the whole wide world. It was no longer a question of exchanging European manufactures – predominantly textiles – for traditional oriental and tropical products, or even of providing outlets for the expanding iron and steel industries by building railways, bridges, and the like. Industry now went out into the world in search of the basic materials without which, in its new forms, it could not exist.

It was a fundamental change, with far-reaching consequences, and it affected every quarter of the globe. The year 1883, for example, saw the discovery and exploitation of the vast Canadian nickel deposits, necessary for the new steel-making processes. By 1900 Chile, which had produced no nitrates thirty years earlier, accounted for three-quarters of the total world production, or 1,400,000 metric tons. In Australia the Mount Morgan copper and gold mine was opened in 1882 and Broken Hill, the largest lead-zinc deposit in the world, the following year. At the same time the demands of the plating and canning industries for tin, and the rapid growth in the use of rubber in the electrical industry and for road-transport, increased the trade of Malaya by very nearly a hundredfold between 1874 and 1914 and made it the richest of all colonial territories. This catalogue could be extended considerably, and it would be necessary in addition to include the stimulus to development in overseas and tropical territories arising from the requirements, already referred to, of growing industrial populations for cheap and plentiful food supplies. The result, in any case, was a transformation of world conditions without parallel in the past. The outer zone of primary producers was expanded from North America, Rumania, and Russia to tropical and subtropical lands and farther afield to Australasia, Argentina, and South Africa; 'areas and lines of commerce that had previously been self-contained dissolved into a single economy on a world scale.'[1] Improvements in shipbuilding, the decline of shipping charges, and the possibility of moving commodities in bulk, brought into existence for the first time in history a world market governed by world prices. By the close of the nineteenth century more of the world was more closely interlocked, economically and financially, than at any time before. In terms of world history – in terms even of European expansion as manifested down to the middle years of the nineteenth century – it was a situation that was entirely new, the product not of slow and continuous development, but of forces released suddenly and with revolutionary effect within the life span of one short generation.

3

It would have been surprising if these new forces had not sought a political outlet. In fact, as is well known, they did. Until a short time ago few historians would have denied that the 'new imperialism', which was so distinctive a feature of the closing decades of the nineteenth century, was a logical expression or consequence of the economic and social developments in the industrialized countries of Europe and in the United States which I have attempted to describe. Latterly, however, there has been a growing

1. *New Cambridge Modern History*, vol. XI (1962), p. 6.

tendency to challenge the validity of this interpretation.[1] 'New, sustained or compelling influences', it has been argued, were lacking in the eighteen-eighties; in particular, the evidence does not indicate that the direction of imperial expansion was influenced to any marked extent by new economic pressures. Some recent writers, indeed, have gone so far as to urge, paradoxically, that the last two decades of the nineteenth century witnessed not a gathering but a slackening of imperial pressures, and that the 'informal' imperialism of the free trade period, though less concerned with political control, had been no less thrusting and aggressive. About these arguments it is sufficient to say three things. The first is that they have done little more, in the last analysis, than replace old conceptual difficulties by new.[2] Secondly, because of their preoccupation with refuting the economic arguments of Hobson and Lenin, they have approached the question from too narrow an angle. And thirdly, by dealing with the problems almost exclusively from a British point of view, they have avoided the main issues. The central fact about the 'new imperialism' is that it was a world-wide movement, in which all the industrialized nations, including the United States and Japan, were involved. If it is approached from the angle of Great Britain, as historians have largely been inclined to do, it is easy to underestimate its force and novelty; for the reactions of Britain, as the greatest existing imperial power, were primarily defensive, its

1. cf. R. Koebner, 'The Concept of Economic Imperialism', *Economic History Review*, 2nd series, vol. II (1949), pp. 1–29; J. Gallagher and R. Robinson, 'The Imperialism of Free Trade', ibid., vol. vi. (1953), pp. 1–15; D. K. Fieldhouse, 'Imperialism: An Hisoriographical Revision', ibid., vol. XIV (1961), pp. 187–209; R. Robinson and J. Gallagher, *Africa and the Victorians* (London, 1961). Other recent interpretations are to be found in R. Pares, 'The Economic Factors in the History of the Empire', *Economic History Review*, vol. VII (1937), pp. 119–44, and A. P. Thornton, *The Imperial Idea and its Enemies* (London, 1959).

2. cf. O. MacDonagh, 'The Anti-Imperialism of Free Trade', *Economic History Review*, 2nd series, vol. XIV (1962), p. 489.

statesmen were reluctant to acquire new territories, and when they did so their purpose was usually either to safeguard existing possessions or to prevent the control of strategic routes passing into the hands of other powers. But this defensive, and in some ways negative, attitude is accounted for by the special circumstances of Great Britain, and was not typical. It was from other powers that the impetus behind the 'new imperialism' came – from powers that calculated that Britain's far-flung empire was the source of its might and that their own new-found industrial strength both entitled them to and necessitated their acquiring a 'place in the sun'.

It is not difficult to demonstrate that the specific arguments of Hobson and Lenin, according to whom imperialism was a struggle for profitable markets of investment, are not borne out by what is known of the flow of capital. That, however, is no reason to suppose that economic motives did not play their part; for the new imperialism was not simply a product of rational calculation, and business interests could be carried away by an optimism which subsequent events disproved.[1] Nor is it difficult to show at any particular point – for example, Gladstone's occupation of Egypt in 1882, or Bismarck's intervention in Africa in 1884 – that the immediate causes of action were strategic or political; but these strategical considerations are only half the story, and it would be hard in the

1. For this reason it is difficult to follow the argument of A. J. Hanna, *European Rule in Africa* (London, 1961), p. 4, who seems to imply that the fact that the chartered company which Rhodes founded in 1889 was 'unable to pay any dividends whatever' until 1923 disproves the generally accepted belief that 'desire for economic gain' was an operative factor in Rhodes's enterprises. In any case failure to pay dividends does not necessarily mean that an undertaking is unprofitable to its promoters. As H. Brunschwig has said, *Mythes et réalités de l'impérialisme colonial français, 1871-1914* (Paris, 1960), p. 106, 'il apparut que des particuliers pouvaient s'enricher aux colonies, même si, du point de vue général ... elles n'étaient pas rentables pour l'état.'

case of Bismarck to deny that it would have been difficult for him to envisage intervention in Africa if it had not been for the new frame of mind in Germany resulting from the rapid industrial development of the Reich after 1871.[1] When we are told that the new imperialism was 'a specifically political phenomenon in origin',[2] the short answer is that in such a context the distinction between politics and economics is unreal. What have to be explained are the factors which distinguished late nineteenth-century imperialism from the imperialism of preceding ages, and this cannot be done without taking account of the basic social and economic changes of the period after 1870. 'I do not exactly know the cause of this sudden revolution,' Lord Salisbury said in 1891, 'but there it is.'[3] His instinctive perception that an abrupt change of mood and temper had occurred, was sound enough. Ever since Disraeli's Crystal Palace speech of 1872, ever since his realization in 1871 that 'a new world' had emerged with 'new influences at work' and 'new and unknown objects and dangers with which to cope', statesmen were conscious of new pressures; and it was these pressures, stemming from the heart of industrial society, that were the explanation of the changed reactions to a power relationship that was substantially older. As the German historian Oncken put it, it was 'as though a completely different dynamic governed the relations of the powers'.[4]

It was only to be expected that the impact of scientific and technological change should take some time to build up. Historians have made much recently of the fact that the doctrine of imperialism was only clearly formulated at 'the very end of the century whose last decades it purported

1. cf. W. Frauendienst, 'Deutsche Weltpolitik', *Die Welt als Geschichte*, vol. XIX (1959), pp. 1–39.

2. Fieldhouse, op. cit., p. 208.

3. Robinson and Gallagher, *Africa and the Victorians*, p. 17.

4. H. Oncken, *Das deutsche Reich und die Vorgeschichte des Weltkrieges*, vol. II (Leipzig, 1933), p. 425.

to interpret',[1] but it would be surprising if it had been otherwise. Theory followed the facts; it was a gloss on developments which men like Chamberlain believed to have been building up over the last twenty or thirty years. In the first place, the industrial revolution had created an enormous differential between the developed and the undeveloped (or, as we would now say, the underdeveloped) parts of the world, and improved communications, technical innovations and new forms of business organization had increased immeasurably the possibilities of exploiting underdeveloped territories. At the same time science and technology had disturbed the existing balance between the more developed states, and the shift which now occurred in their relative strengths – in particular the rising industrial power of imperial Germany and the United States and the gathering speed of industrialization in Russia – was an incitement to the powers to seek compensation and leverage in the wider world. The impact of the prolonged depression between 1873 and 1896 worked in the same direction. Industry was confronted with compelling reasons for seeking new markets, finance for securing safer and more profitable outlets for capital abroad, and the erection of new tariff barriers – in Germany, for example, in 1879, in France in 1892 – increased the pressure for overseas expansion. Even if only a marginal proportion of overseas investment went into colonial territories, the sums involved were by no means negligible, and it is clear that in some at least of the newly acquired tropical dependencies British finance found scope for investment and profit.[2] The position was even

1. Koebner, op. cit., p. 6.
2. This is conceded by Fieldhouse, op. cit., p. 199, who also rightly points out (p. 206) the substantial indirect economic benefits accruing to soldiers, administrators, concession-hunters, and government contractors 'who swarmed in all the new territories'. This aspect of the economics of imperialism has, of course, always been emphasized; cf. Thornton, op. cit., p. 99.

clearer elsewhere – for example, in the Belgian Congo.[1]

From another point of view, the growing dependence of industrialized European societies on overseas supplies for foodstuffs and raw materials was a powerful stimulus to imperialism. Its most conspicuous result was the popularization of 'neo-mercantilist' doctrines. Neo-mercantilism took hold with remarkable speed, first in France and Germany, then in Russia and the United States, and finally in England in the days of Joseph Chamberlain. Since in the new industrial age no nation could hope in the long run to be self-sufficient, it was necessary – according to neo-mercantilist arguments – for each industrial country to develop a colonial empire dependent upon itself, forming a large self-sufficient trading unit, protected if necessary by tariff barriers from outside competition, in which the home country would supply manufactured goods in return for foodstuffs and raw materials. The fallacies inherent in this doctrine have frequently been pointed out, both at the time and subsequently. They did nothing to lessen its psychological impact. 'The day of small nations', said Chamberlain, 'has long passed away; the day of Empires has come.' In many ways the 'new imperialism' reflected an obsession with the magic of size which was the counterpart of the new world of sprawling cities and towering machines.

In the arguments of the neo-mercantilists questions of prestige, economic motivations, and sheer political manoeuvres were all combined and it would be a mistake to try to pick out the one factor or the other and accord it priority. In France, Jules Ferry's speeches reveal a curious mixture of politics, prestige, and crude economic arguments, in which the restoration of France's international standing, depressed by the defeats of 1870 and 1871, loomed

1. Here an investment of fifty million gold francs over a period of thirty years, between 1878 and 1908, produced revenues totalling sixty-six million gold francs by the latter date (Brunschwig, op. cit., p. 71).

large. The same mixture of motives was characteristic of German 'world policy' after 1897, the advocates of which regarded a 'broadening' of Germany's 'economic basis' as essential as a means of ensuring it a leading place in the global constellation which now appeared to be taking the place of the old European balance of power. In the United States, it may be true that the administration was interested primarily in securing naval bases for strategic purposes; but the 'expansionists of 1898' had few doubts or hesitations on the economic score, demanding the Spanish colonies in the interests of trade and surplus capital. As for Russia, economic motives certainly played little, if any, part in the great Russian advance across central Asia between 1858 and 1876 – it would be surprising if at that stage they had done so – but after 1893 the position was different. Witte, the great Russian finance minister, was a convinced and thoroughgoing exponent of neo-mercantilist principles; his monument is the trans-Siberian railway. In the famous memorandum which he addressed to Czar Alexander III in 1892, he set out his ideas on a grand scale. The new railway, Witte said, would not only bring about the opening of Siberia, but would revolutionize world trade, supersede the Suez Canal as the leading route to China, enable Russia to flood the Chinese market with textiles and metal goods, and secure political control of northern China. Strategically, it would strengthen the Russian Pacific fleet and make Russia dominant in Far Eastern waters.[1]

With ideas such as these in the ascendant, it is not surprising that the scramble for colonies gathered pace at an unprecedented rate. By 1900 European civilization overshadowed the earth. In less than one generation one-fifth of the land area of the globe and one-tenth of its inhabitants had been gathered into the imperial domains of the European powers. Africa, a continent four times

1. For Witte's ideas, cf. J. M. Shukow, *Die internationalen Beziehungen im fernen Osten* (Berlin, 1955), p. 50.

the size of Europe, was parcelled out among them. In 1876 not more than one-tenth of Africa was controlled by European powers; during the following decade they laid claim to five million square miles of African territory, containing a population of over sixty millions, and by 1900 nine-tenths of the continent had been brought under European control.

The largest area, some twenty times the size of France, was subjugated by the French, who at the same time were extending and consolidating their position in Tahiti, Tonkin, Tunis, Madagascar, and the New Hebrides. In Asia the French occupation of Annam in 1883, against which Britain reacted by annexing Burma in 1886, opened the assault on the vassal-states of China, and in the last decade of the century all the omens seemed to point to the partition of the Chinese empire itself. France laid claims to the southern provinces of Kwangsi, Yunnan, Kweichow, and Szechwan, comprising a quarter of the total area and nearly a fifth of the population; in reply Britain asserted exclusive interests in the whole basin of the Yangtse, with well over half the total population of the empire, while Russia had set its sights on occupying the vast northern province of Manchuria. Already earlier, within a twenty-years period beginning in 1864, Russia had taken over in central Asia a territory as large as Asia Minor and established for itself 'the most compact colonial empire on earth'.[1] Compared with acquisitions on this scale, the share of imperial Germany was small; but even Germany acquired territories in Africa and the Pacific islands totalling some 1,135,000 square miles and containing a population of thirteen millions.

Last on the scene was the United States, long interested in the Pacific but engrossed, since the civil war, in the opening-up of its own continent. When in the closing years of the century the United States reverted to the

1. O. Hoetzsch, *Grundzüge der Geschichte Russlands* (Stuttgart, 1949), p. 138.

expansionist policies of the 1850s, impelled partly by strategic considerations and partly by fear that the carving out of exclusive spheres of interest in China would be detrimental to its commerce, the triumph of the new imperialism was complete. 'The great nations are rapidly absorbing for their future expansion and their present defence all the waste places of the earth,' wrote Henry Cabot Lodge in 1895; 'as one of the great nations of the world, the United States must not fall out of the line of march.'[1] It did not. The first American move was towards the Hawaiian islands, the annexation of which had been planned during the presidency of Pierce before the civil war. Since 1875 they had been virtually an American protectorate. In 1887 the United States acquired Pearl Harbor as a naval base, and in 1898 it formally annexed the Hawaiian republic. At the same time it declared war on Spain, seized Puerto Rico, Guam, the Mariana islands, and the Philippines, and established a protectorate in Cuba. Few historians would dissent from the view that 1898 was a year of destiny in American foreign relations; it signalized the involvement of the United States in the dialectic of imperialism, in which, after 1885, governments everywhere had been caught up. It was, it seemed, a process in which there was no going back and no standing still, only an inexorable rush forward until the whole world, including even the polar regions which Nansen explored between 1893 and 1896, was brought under the sway of the European conquerors.

There was, without doubt, something febrile and inherently unstable about the 'gaudy empires spatchcocked together' in this way at this time;[2] except for the Russian gains in central Asia few of the territories concerned were destined to remain in undisturbed possession for as much as three-quarters of a century. It was nevertheless a stupendous movement, without parallel in history, which

1. cf. J. W. Pratt, *Expansionists of 1898* (Baltimore, 1936), p. 206.
2. *New Cambridge Modern History*, vol. XI, p. 639.

completely changed the shape of things to come, and to argue, as historians have recently done, that there was 'no break in continuity after 1870' or, still more, that it was an age not of expansion but of 'contraction and decline', does less than justice to its importance. It may be a tenable argument, if we look at the course of development simply from the point of view of causes and origins, that the partition of Africa 'was not the manifestation of some revolutionary urge to empire' but rather 'the climax of a longer process', and that, on the economic side, the late nineteenth-century world was only 'working out, on a much larger scale, the logic of methods inherited from an earlier age'.[1] But if we turn from causes and origins to impact and consequences, the break in continuity and the revolutionary effects of the changes are unmistakable. From the heart of the new industrial societies forces went out which encompassed and transformed the whole world, without respect for persons or for established institutions. Both for the inhabitants of the industrialized nations and for those outside conditions of living changed in fundamental ways; new tensions were set up and new centres of gravity were in process of formation. By the end of the nineteenth century it was evident that the revolution that had started in Europe was a world revolution, that in no sphere, technological, social, or political, could its impetus be checked or restrained. I have dwelt on it at some length, and tried to pick out its main features, because its consequences were so decisive; it was the watershed between modern and contemporary history. In many respects the subsequent chapters will provide little more than a commentary on the effects of the changes that have been surveyed. It is from them that most of the characteristic features of the contemporary world stem.

1. ibid., p. 49.

# III

# THE DWARFING OF EUROPE

*The Significance of the Demographic Factor*

WHEN, at the close of the nineteenth century, the new industrialism reached out from Europe to the four quarters of the globe, it opened an era of change, the consequences of which few contemporaries could even dimly foresee. For most people in Europe the superiority of their values, the irresistible onrush of their civilization at the expense of the 'stagnant' civilizations of the east, were articles of faith; they had no doubt that the spread of empire would quickly result in the dissemination of European civilization throughout the rest of the world. Even Bernard Shaw could argue that, if the Chinese were incapable of establishing conditions in their own country which would promote peaceful commerce and civilized life, it was the duty of the European powers to establish such conditions for them.[1] It was useless to export European skills to backward countries without at the same time introducing European authorities to ensure their proper employment; since the native races were unable to maintain civilized rule themselves, the government of dependencies by the imperial powers was a necessity of the modern world.

This was not simply a question of domination. At one level imperialism might have the appearance of crude, unblushing exploitation; but the leaders of the imperialist movement saw it otherwise. 'In empire', Curzon wrote, 'we have found not merely the key to glory and wealth, but the call to duty and the means of service to mankind'; and Milner saw the British empire as 'a group of states, independent of one another in their local affairs', but

1. cf. A. P. Thornton, *The Imperial Idea and its Enemies*, p. 76.

bound together 'in a permanent organic union' for 'the defence of their common interests and the development of a common civilization'.[1] Joseph Chamberlain's imperial vision, though more specifically economic, was not dissimilar. In his view, the empire would form one 'great commercial republic', an 'economic unit' with its factories in Great Britain and its farms overseas, and a constant flow of population would ensure its prosperity and power.

The ambiguities and inconsistencies in these imperialist designs, particularly the disparity of treatment between the 'white' dominions and the 'coloured' colonies, are not difficult to perceive; but the general pattern, the implicit assumption underlying them all, is clear. What people foresaw was an era in which the European peoples would spread far and wide, settling the new colonial territories, providing in some the bulk of the population, in others at least a strong administrative cadre, but in any event maintaining an indissoluble link with the imperial nexus. This doctrine was explicitly formulated by a governmental committee in 1917. 'The man or woman who leaves Britain', the committee stated, 'is not lost to the Empire, but has gone to be its stay and strength in other Britains overseas.'[2]

It is true that in 1900 the white population of the British empire, some fifty-two millions, was considerably smaller than that of imperial Germany, and of this total scarcely a quarter was living in the empire overseas. But confidence in the maintenance of a flow of emigrants sufficient to clothe the imperial skeleton with flesh and blood was unimpaired. If in the nineteenth century Britain had provided some eighteen million immigrants for the lands of the New World, why should it not maintain a stream of settlers to populate its own colonial empire? Even after

1. ibid., pp. 72, 80.
2. cf. W. K. Hancock, *Survey of British Commonwealth Affairs*, vol. II (London, 1940), p. 128.

the First World War Australian leaders such as Bruce and Dooley were calculating in terms of an Australia with a white population of a hundred millions, and in England Leopold Amery, a member of Milner's imperialist 'kindergarten', saw no reason why, if the United States had expanded in the last century from five to a hundred millions, the British should not grow in the coming century to 'three hundred millions of white people in the Empire'.[1]

It is difficult today to credit the optimistic calculations which the easy successes of the new imperialism fostered at the close of the nineteenth century. The confidence of the European powers in their ability to maintain the position in the world they had won for themselves, their self-assurance and sense of European superiority seem little more than a series of illusions. It was true, of course, that their technical superiority made it easy for them to impose their will by force; this was demonstrated with all possible clarity when they united to suppress the Boxer rising in 1900. The ten years between the Boxer settlement and the downfall of the Manchu dynasty were 'the heyday of western authority in China'.[2] But the maintenance of European superiority by force depended on the existence of at least a relative unity of purpose, which European rivalries tended to cancel out. What would the position be, for example, if one European power deliberately exploited nationalist forces and stirred up rebellion in Asia and Africa in order to weaken its enemies, as Germany did between 1914 and 1918 and Russia after 1917?

During the two or three decades after 1880 practically no one doubted that the European system and the control of the European powers were expanding in an ever-widening circle over the whole of the world's surface. In reality the situation was a good deal more complicated.

1. ibid., p. 149.
2. K. M. Pannikar, *Asia and Western Dominance* (London, 1953), p. 198.

In the first place, in grasping out after possessions and territory in Africa, Asia and the southern seas, the European powers had overreached and overstrained themselves; they had bitten off more than they could chew. Secondly, the interests of the home country and the colonial population rarely ran parallel, and the efforts of the 'white' colonists to control their own affairs provided a precedent or model when the 'coloured' colonial peoples subsequently sought emancipation. Finally, the European powers had set in motion in the extra-European world developments which they could neither halt nor reverse nor control; and these developments were fatal, in the long run, to European predominance. That is why the years of imperialism beginning in 1882 marked at once the apogee and the passing of the European age. Between the Suez crisis of 1882 and the Suez crisis of 1956, the wheel turned full circle; and in the interval the transition from one period of history to another took place.

1

Imperialism itself, in the first place, quickly proved a recalcitrant horse to ride. In England its 'dynamic of self-confident expansion' collapsed almost like a pricked balloon under the strains and stresses of the South African war. Too often, moreover, the results of imperialism proved disconcertingly unlike its promise, just as the profits of empire were often elusive and hard-earned. From this point of view the German colonial empire was a notorious disappointment. Up to 1913 it had absorbed only twenty-four thousand German emigrants but had cost the German taxpayer about £50,000,000; while colonial trade at the same date amounted to only 0·5 per cent of Germany's total commerce. In France, as early as 1899, instead of the looked-for benefits there were loud complaints of the growth of competition from colonial industries and demands for discriminatory tariffs.

But the most disconcerting development of all was the resistance within the empire itself – within the 'white' empire which, in the eyes of imperialists, was the very sinew of the imperial body – to the doctrines of imperialism. The idea of the empire as a 'forever broadening England', 'a vast English nation', 'the all-Saxon home' – the idea which Chamberlain attempted to formulate in institutional terms as imperial federation at the colonial conference of 1897 – appeared to the self-governing dominions, in Sir Keith Hancock's words, as 'a nightmare'. Canada, Australia, New Zealand and later South Africa had no desire for imperial federation, imperial unity, organic links and organic machinery; they were unwilling to surrender their national interests to 'a vast supernationalism claiming for itself the coercive equipment of a sovereign state'.[1]

This was apparent throughout the history of the colonial and imperial conferences from 1887 onwards; it was seen even more directly in the attitude of the 'white' dominions to economic questions and to foreign relations. In both respects they were jealous of their independence; as Sir Joseph Ward, the New Zealand premier, announced in 1911, they would no longer accept the old relation of 'mother and infant'.[2] At the Ottawa conference of 1894, for example, the Australian representatives demanded that they should be freed, like Canada, from the constitutional obstacles which prevented them from instituting differential duties outside the Australasian area; and even earlier, in 1887, New Zealand had claimed the limited rights already enjoyed by Canada of negotiating its own separate commercial treaties with foreign states. None of the 'white' dominions, in short, was willing to forgo the powers essential to economic and political maturity; they were determined, one and all, to liberate themselves 'from the

1. Hancock, op. cit., vol. I (London, 1937), pp. 32, 39.
2. A. B. Keith, *Selected Speeches and Documents on British Colonial Policy, 1763–1917*, vol. II (London, 1933), p. 251.

last vestiges of imperial constraint on their economic freedom'.

The lines of this development are relatively well known and are even clearer in the field of foreign policy than in the economic sphere. From 1882, as France established control in New Caledonia and Germany in Samoa and New Guinea, Australia and New Zealand complained bitterly that the imperial government in London was sacrificing their vital interests in the Pacific to narrow considerations of British policy in Europe; and even earlier, when in 1871 Great Britain appeared to be subordinating Canadian interests to the attainment of a *détente* with the United States, similar complaints were raised in Canada. They were revived and heightened in 1903 by the settlement of the Alaskan boundary dispute on terms which Canadians believed unduly favourable to the United States. 'So long as Canada remains a dependency of the British crown,' the Canadian premier, Sir Wilfrid Laurier, protested, 'the present powers that we have are not sufficient for the maintenance of our rights.'[1]

Already three years earlier, at the time of the Boer War, Laurier had protested at the involvement of the dominions in British military adventures. What 'I claim for Canada,' he said, was that in future she should 'be at liberty to act or not to act, to interfere or not to interfere, to do just as she pleases, and that she shall reserve for herself the right to judge whether or not there is cause for her to act.'[2] This attitude was more moderate than that of the Australian labour leader, William Lane, who said roundly that Australians did not care 'whether Russian civil servants replace the British pauper aristocracy in Hindustan offices'

1. O. D. Skelton, *Life and Letters of Sir Wilfrid Laurier*, vol. II (London, 1922), p. 156.
2. ibid., p. 105; cf. R. M. Dawson, *The Development of Dominion Status, 1900–1936* (London, 1937), pp. 135-6, for a fuller text from the debates of the Canadian House of Commons.

or 'whether the sun sets on the British drum-beat or not – so long as the said drum-beat keeps away from our shores'.[1] But the two attitudes had certain things in common. Negatively, both drew a distinction between British interests and those of the dominions, and refused to subordinate the latter to the former; positively, they issued in a demand for autonomy, particularly in foreign affairs, and for control of the armies and navies on possession of which an independent foreign policy depended. This autonomy was formally denied by Asquith in 1911. The one thing, he insisted, that could not be shared, or decentralized, or delegated, was the foreign policy of the United Kingdom, which was the foreign policy of the British empire.[2] But already before then the principle was in process of being breached. In 1907, for example, Canada had sent emissaries to Japan, thus asserting a place of its own in Pacific politics. In the same year Australia decided to set up a separate naval establishment, serving Australian purposes and subject to Australian control, and three years later Canada followed suit. The corollary was drawn with all finality by the Canadian premier, Sir Robert Borden: 'when Great Britain no longer assumes sole responsibility for defence upon the high seas,' he said, 'she can no longer undertake to assume sole responsibility for and sole control of foreign policy.'[3]

What, of course, crowned these developments and made them irrevocable was the intervention of the dominions in the First World War, their participation in the Peace Conference of 1918–19, and their separate membership of the League of Nations. As Sir Keith Hancock has written, 'it was the challenge of the World War and Canada's response to it' that 'transformed the relationship of Canada

1. cf. M. Bruce, *The Shaping of the Modern World, 1870–1939*, vol. 1 (London, 1958), p. 431.    2. Keith, op. cit., p. 302.
3. Keith, op. cit., p. 309; Dawson, op. cit., p. 162; W. A. Riddell, *Documents on Canadian Foreign Policy, 1917–1939* (Toronto, 1962), p. xliii.

both to the British empire and to the world at large',[1] and what is true of Canada is true of Australia and New Zealand and South Africa as well. The Chanak crisis of 1922, when Lloyd George called upon the dominions to back up British policy by armed force, was a 'melodramatic' intimation of the falsity of calculations of imperial solidarity. The 'Halibut Treaty' between the United States and Canada, signed the following year, was significant not for its contents, which even from a domestic Canadian point of view were of minor importance, but because 'it marked the first occasion that a Canadian foreign minister had negotiated and signed a treaty with a foreign power solely on the authority of his own government.'[2] The decision of Ottawa in 1927 and 1928 to appoint diplomatic representatives in Washington, Paris, and Tokyo was another decisive step on the same road.

By now, however, it was not only a question of asserting (in Mackenzie King's phrase) 'equality of status', but also of putting pressure on London to bring its policy into line with dominion interests. It was under pressure from Canada and South Africa as well as the United States that Britain, in 1921, abandoned its alliance with Japan, and it was under the pressure of Australia and New Zealand that, from 1923, Britain was reluctantly induced to proceed with the fortification of Singapore.

Even so, there were increasing doubts in the Pacific dominions whether the imperial connexion was firm and powerful enough to safeguard their essential interests. As early as 1908 the view was expressed in New Zealand that, although the royal navy was capable of defending either the Atlantic or the Pacific, it was a 'grave question' whether it was equal to both tasks, and both in Australia and in New Zealand, where the resurgence of Japan had given rise to anxiety ever since 1894, people's minds were turning increasingly to the United States as a source of

1. op. cit., vol. I, p. 74.
2. Riddell, op. cit., p. xxv; cf. ibid., pp. 78–87.

support in the event of an emergency in the Pacific.[1] As the First World War loomed nearer, this tendency grew stronger, and in March 1914 it was confirmed by no less a person than Winston Churchill. As First Lord of the Admiralty, Churchill said it was out of the question for him to permit a division of Britain's naval forces in the event of war; consequently Australia and New Zealand must not count on British naval support. If the worst came to the worst, 'the only course of the five millions of white men in the Pacific would be to seek the protection of the United States'.[2]

Thus already before the war of 1914 the course was charted which led to the signature in 1951 of the Pacific Security Agreement between Australia, New Zealand, and the United States, better known as the ANZUS pact.[3] The significance of this treaty in the present context is that Great Britain was not even invited to participate. Already in the previous year the Australian foreign minister had stated that Australians were aware that their destiny was linked 'for all time . . . with the destiny of the United States',[4] and a few years later a Canadian summed up the course of development in the pregnant sentence: 'All roads in the Commonwealth lead to Washington.'[5]

It was an outcome vastly different from that foreseen by imperially minded people in the days of Chamberlain. During the first quarter of the century – in some circles, indeed, down to 1939 and beyond – there had been a general conviction that the 'British genius for compromise'

1. The issues were discussed at some length, for example, in the *Evening Post* (Wellington) on 7 and 8 August 1908; cf. B. K. Gordon, *New Zealand becomes a Pacific Power* (Chicago, 1960), pp. 28–9.

2. Keith, op. cit., p. 353; Gordon, op. cit., p. 30.

3. For the text of the agreement cf. N. Mansergh, *Documents and Speeches on British Commonwealth Affairs, 1931–1952*, vol. II (London, 1953), pp. 1171–3.

4. cf. *International Affairs*, vol. XXVII (1951), p. 157.

5. F. A. Underhill, *The British Commonwealth* (Durham, N.C., 1956), pp. xviii, 99.

would find a 'middle way' which would satisfy at one and the same time the aspirations of the dominions for autonomy and the maintenance of imperial unity; and for something like a generation it seemed as though this confidence had been vindicated by the Balfour report of 1926 and the Statute of Westminster of 1931. Today these two documents are a good deal less impressive than they appeared a quarter of a century ago to commentators bewitched by the subtle alchemy which (they believed) was enabling the British empire to find a way out of dilemmas to which other imperialisms had succumbed. The reason, of course, is that the problem to be solved was not merely an exercise in political theory. It was not simply a question of devising a constitutional formula which would substitute 'influence' for direct government and reconcile the concepts of *imperium* and liberty, but, far more fundamentally, it was a matter of keeping up with the facts of a world revolutionized by the explosiveness of new scientific knowledge, the new technology and the new imperialism. No doubt, the 'white' dominions had national aspirations of their own; but it would be very misleading to picture them as straining to escape from the imperial leash. On the contrary, they were loyal to the imperial design, as their prompt response in the wars of 1914–18 and 1939–45 amply demonstrated. But they, like Great Britain, were caught in a vice, constrained by hard facts from which there was no escape. In the long run, it was not their wishes or their attitude to the imperial connexion that counted, but revolutionary developments in the world around them to which they were forced to react, often in ways that ran contrary to what, left to themselves, they would have freely chosen.

It was these pressures that were the decisive factor, rather than the constitutional changes leading from Empire to Commonwealth which have usually been picked out for comment. Not the Statute of Westminster but the ANZUS pact had in it the seeds of the future. It was a paradox of the new imperialism that it released pressures which made its own tenets unworkable. By stirring the outer world into activity it loosened the ties of empire, just as it undermined the pre-eminent position of Europe, which was its most cherished belief. In every decade after 1900 it became more clear to more people that future centres of population and power were building up outside Europe, that the days of European predominance were numbered, and that a great turning-point had been reached and passed. A new world was in the making.

2

For many people the most striking aspect of the change was the rising importance of the United States of America. It was no accident that the influential English journalist, W. T. Stead, penned his widely read pamphlet, *The Americanization of the World, or the Trend of the Twentieth Century*, in 1902. Already in 1898 he had written to Lord Morley: 'I feel as if the centre of the English-speaking world were shifting westwards,' and four years earlier Conan Doyle, at that time in America, had remarked: 'The centre of gravity of race is over here, and we have to readjust ourselves.'[1]

But the emergence of the United States as (in Stead's words) 'the greatest of world-powers' was only one aspect of a far more extensive process. Already earlier farsighted people had perceived the existence of trends that were making the Pacific 'an ocean of destiny'. In America Lincoln's secretary of state, Seward, was alive to its potentialities for the United States; in Europe the brilliant Russian exile, Alexander Herzen, described the Pacific succinctly as the 'Mediterranean of the Future'.[2] But it was

1. cf. R. H. Heindel, *The American Impact on Great Britain, 1898–1914* (Philadelphia, 1940), pp. 53, 130–1.
2. cf. M. Laserson, *The American Impact on Russia, 1784–1917* (ed. 1962), p. 270.

the events of the last decade of the nineteenth century – the Japanese attack on China in 1894, the American seizure of the Philippine islands in 1898 – that for most people brought Pacific affairs into sudden prominence. 'The power that rules the Pacific', declaimed Senator Albert J. Beveridge, 'is the power that rules the world';[1] and his words were promptly taken up and echoed by President Theodore Roosevelt. 'The Mediterranean era died with the discovery of America,' said Roosevelt; 'the Atlantic era is now at the height of its development, and must soon exhaust the resources at its command; the Pacific era, destined to be the greatest of all, is just at its dawn.'[2]

These bold predictions reflected the speculations of American philosophical imperialists, such as Alfred Thayer Mahan and Brooks Adams; but they also expressed the primacy which Asia and the Pacific had acquired in American thought on foreign affairs. There were solid reasons for this change of emphasis. By 1900 it was impossible to ignore the fact that a major shift in the world's population was beginning to take place and that the demographic balance was turning against Europe.[3] The

1. cf. R. W. van Alstyne, *The Rising American Empire* (Oxford, 1960), p. 187.

2. cf. A. C. Coolidge, *The United States as a World Power* (New York, 1908), p. 325. It was a theme to which Roosevelt constantly reverted. 'The empire that shifted from the Mediterranean', he said at San Francisco in a speech on 13 May 1903, 'will in the lifetime of those now children bid fair to shift once more westward to the Pacific'; *Californian Addresses* (San Francisco, 1903), p. 96.

3. Probably the best general historical source for population figures and population changes is E. Kirsten, E. W. Buchholz, and W. Köllmann, *Raum und Bevölkerung in der Weltgeschichte* (generally referred to as *Bevölkerungs – Ploetz*); the second volume *(neuere und neueste Zeit*, 2nd ed., Würzburg, 1956) contains the relevant data. It should be supplemented for more recent years by the *Demographic Year Book of the United Nations* (the most recent edition of which, at the time of writing, was the 13th issue, New York, 1961), the annual volumes of the *Statesman's Year Book*, etc. Among the literature there is still no general work to supersede A. M. Carr-Saunders,

expansion of Europe in the nineteenth century had been based on a phenomenal population growth which doubled the population of the continent, enabled it at the same time to export forty million emigrants, and even so raised its proportion of the total world population, relative to other continents, from one-fifth to one-quarter. By 1900 this rate of growth was visibly slackening. From about 1890 – which thus, once again, stands out as an important turning-point – there was a steady decline in the birth rate in industrialized Europe, the consequence not only of the spread of contraception but also of a higher standard of living and in some degree, perhaps, of the long economic depression. For the next two decades it was masked, except in the case of France, by an even steeper decline in the death rate; but from approximately 1890 in England, from 1905 in Switzerland, and from 1910 in Germany, a decline in the net reproduction rate set in, until in the decade 1920–30 it fell below unity. From this trend Russia, with a rate of increase of 1·7 per cent, was significantly an exception; but by 1930 both the United States, which at the beginning of the century had a high net increase, and the 'white' colonies overseas, particularly Australia and New Zealand, had fallen into line with industrial Europe.[1]

Contrasted with this tendency for the population of Europe and the 'white' territories overseas to fall, there was, on the other hand, an equally dramatic tendency for that of the peoples of Asia and Africa to rise. This was directly 'a result of the imperial policies' practised during the last decades of the ninteenth century.[2] More

*World Population* (London, 1936). W. D. Borrie, *Population Trends and Policies* (Sydney, 1948), and W. S. Thompson, *Population and Peace in the Pacific* (Chicago, 1946), are, however, particularly useful for the areas concerned.

1. Borrie, op. cit., pp. 66 f. In the U.S.A. the net rate of reproduction at no time exceeded 0.98 between 1930 and 1940, i.e. it was below the replacement rate; cf. M. A. Reinhard, *Histoire de la population mondiale* (Paris, 1940), p. 373.

2. Borrie, op. cit., p. 240; cf. Thompson, op. cit., pp. 299–301.

specifically, it was a consequence of the maor advances in hygiene and medicine which were so notable a feature of the period, of the introduction of improved techniques of agriculture, which offset the effects of intermittent famine, of irrigation, land reclamation, and improvements in transport and food storage facilities. Thus the population of India, though fluctuating violently until 1920, grew rapidly thereafter, the increase in the subsequent twenty years (eighty-three million) being equivalent to two-thirds of the total population of the United States at that time. In Japan, where the population appears to have been stable in the century and a half prior to 1872, the subsequent sixty years saw a steady, but by no means phenomenal growth, chiefly as a result of a reduction in the death rate, and between 1872 and 1930 the population doubled.[1]

In the case of China, as in that of Africa, lack of accurate evidence makes all estimates of doubtful value. In both instances it has usually been assumed that the population remained static 'at or near the Malthusian limit', or even that in China it actually declined, as a result of pestilence and famine, during the nineteenth century.[2] But the consideration that mattered was less what the population was than what it would be, once the introduction of measures of hygiene and other forms of modernization got under way, as they did in China after the end of the civil war in 1949. The population of mainland China, as recorded in the census of 1953 – namely, five hundred and eighty-three millions – was far ahead of the most liberal estimate accepted up to that time – namely, four hundred and fifty millions – and implied, with a net annual rate of increase of 2·8 per cent, a growth rate of over sixteen million a

1. The Japanese population growth is examined in some detail by Thompson, op. cit., pp. 93–175; cf. also Ryoichii Ishii, *Population Pressure and Economic Life in Japan* (London, 1937).

2. Carr-Saunders, op. cit., pp. 34–42, 286–90. Carr-Saunders believed that the absolute increase in Africa between 1900 and 1923 may have been as much as twenty-five million (p. 35).

year, or an accretion in a ten-year period equivalent to considerably more than the total population of the United States in 1953.

3

From these demographic trends two conclusions seemed to follow. The first was that a dramatic shift was taking place in the balance between the 'white' and 'coloured' races; the second was that the differential rate of population growth, in conjunction with migration and population movements on a continental scale, was leading to the formation, far away from Europe, of new centres of population, production, and power.

Of the two developments it was the former that was the first to attract widespread attention. In reality, the rate of population growth in Asia in the first quarter of the twentieth century was far from exceptional, and Europe not only retained but actually increased its proportion of the total estimated world population between 1850 and 1930. Even the growth per cent of the population of Japan between 1870 and 1920 was smaller than that of England and Wales and of Russia during the same period. What impressed contemporaries, however, was not the rate of growth itself, but the density of population in Asia and consequently the marked effect in total increase which only a small improvement in the net rate of reproduction might have.

In Japan, for example, the average population density around 1930 was four hundred and thirty-nine per square mile, as compared with an average density in Europe (excluding the Soviet Union) of one hundred and eighty-four, in the Soviet Union of a little more than twenty, and in the habitable areas of Australia of 3.8 per square mile. Since, however, the northern island of Hokkaido, snow-bound for five months in the year, is infertile and inhospitable, the average figure of four hundred and thirty-nine

per square mile is misleading, and on the Honshu mainland the comparable figure was, in fact, five hundred and
fifty-three. Even so, the density of population in Japan
was far below that in Java where, as a result of a phenomenal increase since 1850, the density had reached eight
hundred and seventeen per square mile by 1930; while in
the Kiangsu province of China it was probably not short
of a thousand and in the plain of Chengtu was thought to
reach one thousand seven hundred.[1]

The significant fact arising from these figures was that
they appeared to imply a pressure on resources, at or about
subsistence level, the only remedy for which, failing restric-
tion of the birth rate, was migration from the 'over-
populated' to the 'underpopulated' continents. How otherwise was it possible to envisage coping with an annual
population increase which, for China, India, and Japan
alone, on figures calculated at the beginning of the twentieth century, could hardly be smaller than seven and a
half million and would probably be a good deal larger?
The result of the impact of the west, it appeared, had been
to induce the first phase of the demographic cycle (viz.
decreasing mortality) but not the second (viz. declining
fertility), and in spite of the example of Japan industrialization was as yet scarcely conceived of as a solution of the
'Malthusian dilemma'.

As, from about 1900, the significance of the falling European birth rate and the likely consequences of declining
mortality in Asia became matters of general knowledge,
there grew up, among Europeans and descendants of
European stock overseas, an almost neurotic awareness of
the precariousness of their position in the face of an
expansive Asia. In the end, they began to ask themselves,
how could they hope to avoid being borne down by the
sheer weight of numbers? It was, perhaps, the first sign of

1. Carr-Saunders, op. cit., p. 287, gives the figure of 900 per square
mile for Kiangsu, but in view of the later census figures this estimate
is probably low; the density today is around 1,150.

inner misgiving, of an instinctive realization that, without their knowing it, their intervention in Asia and Africa had set in motion undercurrents which, when they came to the surface, would sweep the stream of world history into new channels.

4

One of the first signs of the demographic crisis was the shrill cry of the 'Yellow Peril', taken up by the German emperor at the time of the Boxer rising, and stimulated by the victory of Japan over Russia in 1905.[1] It gave rise quickly to a semi-popular and largely sensational literature, of which Sir Leo Chiozza Money's *The Peril of the White* (1925), with its message 'Renew or Die!' may be regarded as a characteristic example. The theme of all these books was the precariousness of the European position in the world, when, as Money pointed out, there were only three hundred and four thousand British in Asia in an aggregate population of three hundred and thirty-four millions (or less than one per thousand) and only seven thousand four hundred Europeans in British West Africa out of a population of nearly twenty-three millions. India was the heart of the empire; but in the extensive Dacca and Chittagong divisions of Bengal, with a population of seventeen and a half millions, there were in 1907 only twenty-one British covenanted civil servants and twelve British police officers.[2]

But it was, above all, on Australia and New Zealand that attention concentrated, for here – as one writer said of Australia in 1917 – 'a population less than the depleted population of Scotland' was 'pathetically struggling to hold a continent as a white man's land against the

1. The fullest and most able account of the 'Yellow Peril' scare, in its international ramifications, is in H. Gollwitzer, *Die gelbe Gefahr* (Göttingen, 1962).

2. cf. *Cambridge History of the British Empire*, vol. v (1932), p. 252.

congested millions of coloured peoples just across the sea'.[1] We may smile at the extravagance of language which was a feature of writing of this type; but on a more sober level there was plenty of evidence of the reality of demographic pressures coming from Asia. China, in particular, which even so cautious an observer as Carr-Saunders likened to a 'saturated sponge',[2] accounted for a vast and continuous outflow of emigrants. Of these the largest proportion – in some years more than a million – went to Manchuria, where they came up against the stream of Russian colonization from the west; but a substantial number – between 1920 and 1940 perhaps two millions – settled in south-east Asia, where they added to the pressure of the already dense population bearing down on Australia and New Zealand.

The immediate response of the countries concerned was to erect a ring fence of stringent immigration laws and regulations so framed as to exclude non-Europeans. In Australia the federal government, building on earlier enactments of the individual states, introduced legislation in 1901 to enforce a rigorous 'white Australia' policy; and similar measures were put into force in New Zealand, Canada, and the United States. Their effectiveness is beyond doubt. But for these restrictions, as Carr-Saunders observed in 1936, it seems almost certain that, by that date, 'the population of the western seaboard of North America would have been largely Asiatic'.[3] But no one supposed that exclusion was a long-term solution, and already in 1904 a royal commission set up to investigate the decline of the birth rate in New South Wales expressed the view that, failing a high birth rate and large-scale 'white' immigration, it was necessary to envisage the possibility that 'Australia might be lost to the British'.[4]

From that time forward the 'threat' of the 'teeming

1. James Marchant, *Birth Rate and Empire* (London, 1917), p. 3.
2. Carr-Saunders, *World Population*, p. 294.
3. ibid., p. 190.
4. Borrie, op. cit., p. 58.

millions of Asia' to Australia and other colonies under European control, and the measures necessary to counter it, became a constant theme, debated recurrently and inconclusively by committee after committee and commission after commission, to the accompaniment of pessimistic prognostications from 'experts' and publicists.[1] It would be otiose to follow the course of the debate,[2] for all it revealed was an insoluble dilemma which was itself the unhappy and unawaited 'end product of the imperialism practised by the western powers in the nineteenth century'.[3] The only hope of maintaining the imperial position and pre-eminence of the European powers overseas (the argument ran) was to keep up the volume of emigration. But, except for Italy, none of the western colonial countries – not even Nazi Germany – could boast a replacement birth rate. On the contrary, they themselves were faced by a manpower crisis, and it was entirely unrealistic to look to them for a flow of emigrants. There was only one conclusion: 'the maintenance of the barriers depended ultimately upon the distribution of armed power.'[4]

But could Europe be relied upon to provide the armed power necessary for the defence of its overseas territories? The grim reality seemed to point in the opposite direction. As the imperial powers which had hitherto exercised

1. Thus Sir Leo Money (op. cit., p. 83), while deprecating the alarmist contention that 'the rifle or machine gun or aeroplane or "chemicals" may arm a Yellow attack upon the west', nevertheless held that 'the possibility of Europe perishing through the employment by the coloured races of its own scientific methods of destruction cannot be wholly dismissed'. He did not, however, believe that the real danger lay in this direction, for in his view there was no need for 'weapons to destroy European life and civilization if the Whites in Europe and elsewhere are set upon race suicide'. This pessimism became a continuing theme. 'Unless Australia doubles her population,' R. G. Casey prognosticated in 1951, 'in a generation our children will be pulling rickshaws'; *International Affairs*, vol. XXVII (1951), p. 200.

2. It is summarized by Hancock, op. cit., vol. II, i. pp. 149–77.

3. Borrie, op. cit., p. 30.

4. Hancock, op. cit., p. 177.

control in the dependent areas of Asia and Africa were confronted by an absolute decline in their numbers, two things became clear: first, that there was a narrow limit to the reserves of manpower available for normal peace-time policing and administration, and, secondly, that their capacity to defend these areas against internal revolt and attack from without was being eroded. No doubt it needed more than erosion to bring the established imperial structure crashing down; but when the Second World War shattered the flimsy international equilibrium upon which its maintenance depended, the demographic factor came into its own. The collapse of the European empires in Asia in 1941 was essentially a demographic failure; and there is no doubt that demographic factors played their part in the following years in bringing about the withdrawal of the British from India, the Dutch from Indonesia, and the French from Indo-China.[1]

5

The 'differentials in population growth', it has been said, were working against Europe and in favour of Asia.[2] One conclusion frequently drawn was that 'western Europeans' were 'going to be on the losing end of power arguments in the near future'. But, apart from such prognostications, which might or might not be true, the facts were eloquent enough. As Mackenzie King said in 1939, there had been a 'change in the world balance of power', a 'change in strategic conditions', a 'change in economic needs and in relative industrial capacity', and he went on to draw the conclusion that, in view of the shift in 'the balance of world power', it was necessary for Canada 'to keep its Pacific as well as its Atlantic coast in mind'.[3]

More than a decade earlier Mackenzie King had drawn

1. Borrie, op. cit., p. 29.     2. Thompson, op. cit., p. 341.
3. Riddell, op. cit., p. 219.

attention to the fact that Canadian 'trade with the Orient today is greater than was the trade of Canada with the United Kingdom at the time the government of Sir Wilfrid Laurier came into office', i.e. in 1896.[1] These were facts of great importance and wide application. What Mackenzie King said of Canada applied perhaps even more forcibly to Australia. The Australians also were forced to 'the realization that they constitute primarily a Pacific nation', and that 'every question of national policy' must be considered in a Pacific context.[2] And once the completion of the great transcontinental railroads had linked New York with San Francisco, there was, as has already been noted,[3] a similar shift in the economic and demographic axis of the United States; as the great movement of westward expansion got under way, not only was the attachment of the United States to the Atlantic economy weakened, but American policy also took on a specifically Pacific orientation.

It is no accident that parallel developments were visible at the same time in Russia. Nothing, perhaps, is more significant in Russian history at this period than the new emphasis on Asia, not as an empire of nomadic tribes to be conquered and ruled, but as territory to be settled and developed. For Dostoyevsky Asia was Russia's 'undiscovered America'; there, rather than in Europe, he said, lay Russia's hopes for the future.[4] In fact, the rapid rise of population following the emancipation of the peasants in 1861 provided a vast reservoir of manpower for colonization, and from 1881 official measures were taken to encourage emigration eastwards. The result was the creation of new centres east of the Urals, parallel in some respects to the American middle-western industrial centres in Illinois, Michigan, and Ohio, and the movement of a vast

1. ibid., p. 281; cf. pp. 286–7.    2. Borrie, op. cit., p. 236.
3. Above, pp. 42–3.
4. cf. *The Diary of a Writer* (trs. Boris Brasol, London, 1949), p. 1048.

stream of settlers into Siberia, parallel to the migration in the United States across the Rockies into California.

The colonization of Siberia and of California were major demographic events. In both regions the population growth was stupendous, and in both it was essentially a twentieth-century phenomenon. In California, where even after the gold rush of 1848 and 1849 the population was still under a hundred thousand, it had reached almost a million and a half in 1900, i.e. a fifteenfold increase in fifty years. But this was only a start, and the real leap forward came later. Between 1920 and 1940 the population more than doubled from 3·4 to 6·9 millions; between 1940 and 1960 it rose at a fantastic pace to 15·7 millions, until finally in 1963 California became the most populous state in the Union.

In Siberia and Russian Asia the demographic pattern, though less spectacular, was curiously similar. Here also the nineteenth century saw a substantial movement of emigrants, amounting in all probability to around four millions;[1] but, here again, the major developments occurred in the following century. In the first place, the pace quickened after the completion of the trans-Siberian railway, and between 1900 and 1914 probably another 3·5 million emigrants left for the lands beyond the Urals. It was, however, after the setbacks of war and revolution that the really large-scale development of the lands beyond the Urals was undertaken. From 1929 onwards the settlement and industrialization of Soviet Asia became a major objective of Soviet planning, and the result was a rapid increase in the population of the Asiatic territories, which was later intensified by the redeployment of industry east of the Urals after the outbreak of war with Germany in 1941. Between 1926 and 1939 the population of Soviet Asia rose

1. Estimates vary. Carr-Saunders (op. cit., p. 56) gives 3·7 millions for 1800–1900; other estimates are as high as 4·5 millions. Calculations are complicated by returning emigrants and by the fact that the figure for registered emigrants is certainly an underestimate.

by approximately ten millions; after 1939 the eastward trend was accentuated. Thus while, as a result of the losses inflicted during the war, the total population increase of the Soviet Union between 1939 and 1959 was only 9.5 per cent, that of central Asia and Kazakhstan rose by 38 per cent, of eastern Siberia by 34 per cent, and of the Far East province by no less than 70 per cent, or from 2.3 to 4 millions.

6

This great demographic revolution was also an economic revolution. Both in Soviet Asia and in the American west the shift in population was accompanied by a shift in industry and a rise of new industrial centres. The population of Los Angeles rose from 102,000 in 1900 to almost 2.5 millions in 1960, that of San Francisco which had developed earlier, from 342,000 to 740,000. In Soviet Asia the process of urbanization was no less intense. Novosibirsk (formerly Novonikolaievsk), the capital of western Siberia, had 5,000 inhabitants in 1896; by 1959 the number had risen to 887,000. Magnitogorsk, which had only thirty-seven families of semi-nomadic herdsmen in 1926, had 145,000 inhabitants by 1939 and 284,000 in 1956. According to the official statistics industrial production increased 277 per cent in central Asia and 285 per cent in Siberia during the decade following the introduction of the first five-year plan in 1928; indeed, as early as 1935 a German economist asserted that the Soviet government was only able to carry through its programme of 'socialism in one country' through the creation of new industrial centres in the east.[1]

In effect the demographic changes in the United States and the Soviet Union signified a shift from the Atlantic to the Pacific seaboard. Their implications were reinforced by parallel changes elsewhere, particularly by the rising

1. P. Berkenkopf, *Siberien als Zukunftsland der Industrie* (Stuttgart, 1935), p. 10.

importance of the southern hemisphere which had hither-to supported only an insignificant fraction of the popula-tion of the globe. Australia and New Zealand, with only one and a quarter million inhabitants in 1860, had over twelve and a half millions a century later. But the most striking increase was in South and Central America. Here the population was some twenty-six millions behind that of North America in 1920; but by 1960 it had overtaken it by seven millions, although the population of North America had increased from 117 to 199 millions in the intervening period.

Here again a great new centre of population was arising, far from Europe and outside the European sphere of interest. The growth of the population of the Argentine and Brazil was phenomenal, particularly in the twentieth century. In the Argentine a population of only three-quarters of a million in 1850 increased more than sixfold to 4·9 millions in 1900. By the First World War another four millions had been added; by 1920 the population had reached ten millions, and by 1960 it had doubled again to over twenty millions. Brazil, with about five and a half million inhabitants in 1850, increased its population over threefold to seventeen millions in 1900, and then soared ahead; by 1930 its population had almost doubled at thirty-three millions, and in 1960 it had more than doubled again at more than seventy millions.

The significance of this process of redistribution is emphasized if we consider, in addition, the progress of urbanization. In 1900, as was noted earlier,[1] there were fourteen cities with a population of one million or more, and of these six (including St Petersburg and Moscow) were in Europe, three in Asia, three in North America, and two in South America. By 1960, when the total had risen to sixty-nine, the distribution had changed radically. No less than twenty-six (i.e. more than thirty-seven per cent) were in Asia; the number in Latin America had risen to

1. Above, p. 53.

eight (as against seven in the United States and Canada) and three further cities of over a million, two in Australia and one in South Africa, indicated the rising importance of the southern hemisphere. Of twenty-eight cities with more than two million inhabitants in 1960, five were in Europe (excluding European Russia), eleven in Asia, four in North America, four in Latin America, and two in the Soviet Union.

**DISTRIBUTION OF CITIES ABOVE ONE MILLION INHABITANTS**

| 1960 | Above 2 million Inhabitants | 1–2 million Inhabitants | Total |
|---|---|---|---|
| Asia: Far East (China, Japan, Korea, Philippines) | 7 | 9 | 16 |
| Asia: South and South East (India, Pakistan, Thailand, Indonesia) | 4 | 6 | 10 |
| Europe (excluding U.S.S.R.) | 5 | 14 | 19 |
| U.S.S.R. (Europe and Asia) | 2 | 1 | 3 |
| North America (U.S.A., Canada) | 4 | 3 | 7 |
| South and Central America | 4 | 4 | 8 |
| Middle East (Egypt, Persia, Turkey) | 1 | 2 | 3 |
| South Africa | — | 1 | 1 |
| Australia | 1 | 1 | 2 |
| | 28 | 41 | 69 |

These figures are, of course, arbitrary in regard to the process of urbanization as a whole. In particular, they do less than justice to the progress of urban development in Soviet Asia, where there was a steep rise in the number of cities in the 250–500,000 category.[1] Between 1926 and 1939

1. There are useful lists in *Bevölkerungs-Ploetz*, pp. 342–3 and 345–6.

the cities of the Soviet Union more than doubled their population, a rate of increase which took the United States about thirty years and most European countries little short of a century; and between 1939 and 1959 the urban population rose further from thirty-two to forty-eight per cent of the total. Indeed, the urban revolution in the Soviet Union and subsequently in China took place at a speed greater than ever witnessed before, and its result was to intensify the existing tendency for the economic centre of gravity to shift away from western Europe.

The share of Europe in the total estimated world

|  | 1920 | | 1950 | | 1960 | |
|---|---|---|---|---|---|---|
|  | millions | per cent | millions | per cent | millions | per cent |
| Europe | 328 | 18·1 | 393 | 15·9 | 427 | 14·3 |
| U.S.S.R. | 158 | 8·7 | 181 | 7·3 | 214 | 7·2 |
| Asia | 967 | 53·5 | 1,360 | 55·0 | 1,679 | 56·0 |
| Africa | 140 | 7·7 | 199 | 8·0 | 254 | 8·5 |
| North America | 117 | 6·5 | 168 | 6·8 | 199 | 6·6 |
| Central and South America | 91 | 5·0 | 162 | 6·5 | 206 | 6·9 |
| Australasia and Oceania | 9 | 0·5 | 13 | 0·5 | 16 | 0·5 |
| Total World Population | 1,810 | 100 | 2,476 | 100 | 2,995 | 100 |

population, which had risen by five per cent between 1850 and 1913, declined (as the above table shows) by 3·8 per cent between 1920 and 1960, and it is noteworthy that the decline gathered pace in the decade 1950–60. The significance of this population trend was that it indicated not only the operation of differentials in population growth which were working against Europe, but also the rising importance of non-European centres of productivity, civilization and ultimately of power. The conclusion to which these facts seemed to lead was pointed out by an

American historian as long ago as 1943.[1] The centre of gravity, he wrote, was shifting 'to countries outside of Europe'. This shift, which started at the turn of the nineteenth and twentieth centuries, was a consequence of the imperialism which characterized the new industrial age beginning around 1870. At first the spread of European power and technology seemed to mean 'pushing frontiers farther away from a centre which became stronger and stronger as it ruled a wider area'. But 'imperceptibly this development changed its character'. The centre itself was 'dislocated and transferred to other continents', and, stimulated by European capital, European inventions, European manpower and European standards of living, new non-European and extra-European centres came into existence.

There is an irony about this process which it is impossible to overlook. The Chinese and Japanese in the middle of the nineteenth century asked only to be allowed to avoid all possible contact with the outside world and to live on their own resources in the traditional manner. The western powers forced them to open their countries to western penetration, and in this way set in motion demographic movements which they could not reverse. It is true, of course, that the new demographic trends took time to work themselves out. But after fifty years it was evident that the European powers, far from creating (as most people had expected) a world in their own image, had raised up forces in Asia and Africa which would not rest content until they had challenged the political hegemony of Europe.

Demographic factors alone are a precondition rather than a cause of political power, and the significance of mere size is often disputed. Nevertheless it is clear enough that, if Great Britain had remained a country of twelve million inhabitants, as it was in 1801, not even a high degree of industrialization would have enabled it to attain

1. E. Fischer, *The Passing of the European Age* (Cambridge, Mass., 1943), p. xii.

the dominant position it achieved in the latter half of the nineteenth century. During the century between 1815 and 1914 differentials in population were neutralized to a considerable extent by differentials in industrial capacity, and industrialized countries like England and France were able to gain control over far larger populations in non-industrialized lands with comparative ease. But this predominance, which seemed a permanent dispensation, was in reality only a temporary factor, for it was soon evident that technical skills were not the monopoly of any part of the world and that it was easy to transfer them from one country to another. It was also demonstrated, both by Soviet Russia and by China, that in case of urgent need capital could be accumulated at great speed – though also at great human cost.

Hence, as the twentieth century proceeded, the advantages which had ensured European predominance – namely, the monopoly of machine production and the military strength conferred by industrialization – receded and the underlying demographic factors reasserted their importance. It is no exaggeration to state that the demographic revolution of the half-century between 1890 and 1940 was the basic change marking the transition from one era of history to another. At the same time the period of European political hegemony was drawing to a close and the European balance of power, which for so long had governed the relations between states, was being superseded by the age of world politics.

# IV

# FROM THE EUROPEAN BALANCE
# OF POWER TO THE AGE OF
# WORLD POLITICS

*The Changing Environment of International Relations*

FOR anyone looking down on the world of 1960 and comparing it with the world of 1870 or 1880, nothing will probably be more striking than the change which has taken place in the structure of international relations. Seventy-five years ago the predominance of the European powers was unchallenged; the political map of Asia and Africa was drawn by statesmen in London, Paris, and Berlin, and the Russian general, Dragomirov, was only echoing the belief of his day when he haughtily announced that 'Far Eastern affairs are decided in Europe'.[1] Today this has long ago ceased to be even approximately true. By the end of the Second World War the collapse of the old system of balance of power was evident for all to see, and it was also clear that its collapse was not simply the result of the war, but rather the consequence of a far longer process of erosion which abnormal circumstances – the isolationism of the United States after 1919 and the weakening of Soviet Russia by revolution and civil war – had masked but not halted. The structure of great power politics, and its modalities, in the age of Khrushchev were essentially different from those of the age of Bismarck. Instead of by a concert of powers, we were confronted by two great super-powers, the Soviet Union and the United States, whose pre-eminence was based on their quasi-monopoly of nuclear weapons and of the delivery-

1. cf. G. F. Hudson, *The Far East in World Politics* (2nd ed., London, 1939), p. 74.

systems for launching nuclear weapons; and although Russia has one foot in Europe, it is significant that both superpowers are great continent-spanning federal states, neither of which can realistically be classed as European. Thus in the space of half a century a multilateral system of equilibrium centred upon Europe had been displaced by a system of global bipolarity between the two great extra-European powers, the United States, and the Soviet Union.[1]

1

There can be no doubt that the change in Europe's political standing was a revolution of first magnitude, which radically altered the character and conditions of international relations. How did it come about?

The most obvious explanation, and the one usually emphasized, is the exhaustion of Europe in two world wars. Coupled with this was its fragmentation into an increasing number of small and medium-sized units, which culminated at the peace settlement of 1919 in the creation of the succession-states of eastern Europe – states which were too

1. There has been considerable discussion in the English-speaking world of bipolarity as a problem of contemporary politics, but very little has yet been done on the historical side to elucidate its origins. The first attempt to deal with the question in general terms was Ludwig Dehio's *Gleichgewicht oder Hegemonie* (Krefeld, 1948), the arguments of which I briefly summarized for English readers in *History in a Changing World* (Oxford, 1955), pp. 168–84; Dehio's book has subsequently been translated under the title *The Precarious Balance. The Politics of Power in Europe, 1494–1945* (London, 1963). Since then the problem has been discussed from a number of different angles by E. Hölzle. Among his writings on the subject the following will probably be found the most useful: *Geschichte der zweigeteilten Welt* (Hamburg, 1961); *Die Revolution der zweigeteilten Welt* (Hamburg, 1963); 'Das Ende des europäischen Staatensystems', *Archiv für Kulturgeschichte* XL (1958), pp. 346–68; 'Der Dualismus der heutigen Weltreiche als geschichtliches Problem', *Historische Zeitschrift* CLXXXVIII (1959), pp. 566–93.

weak, too small and too divided to maintain their independence. Every European war, it has been pointed out,[1]
resulted in greater division, every colonial war in greater
cohesion. But, though the self-destruction of Europe
through its internecine struggles is usually singled out as
the main cause of its political decline, over the long run
two other factors were more decisive. The first was the
concentration of power on the two flanks – a process in
most respects independent of what went on in Europe.
The second was the rise of new centres of political gravity
and new fields of conflict in Asia and in the Pacific region,
in which the European powers were only obliquely concerned.

To regard the division of the world into two great power
*blocs* as simply a result of the decline of Europe is too
negative an interpretation. Long before there could be any
question of the decline of Europe – indeed, at the time
when its energies were actually developing their maximum
thrust – international politics were breaking through their
European setting. From early in the nineteenth century,
from the time of the independence of Latin America, it is
possible to distinguish two spheres of policy, overlapping
but distinct, the one the familiar conflict of the European
powers, the other the greater global conflict on a world-
wide scale, in which, as first Spain and then France fell
behind, Great Britain, Russia, and the United States
emerged as protagonists. The advent of the era of world
politics meant that new interests were at play and that old
interests were viewed in a new light; the traditional
objectives of European policy were no longer the only, not
necessarily even the main, criterion. Of this there is perhaps
no clearer illustration than the agreement made in 1915
to hand over Constantinople to Russia – an agreement
which reversed what for a century had been regarded as a
basic principle of European policy. In the last analysis the

1. cf. Dehio, *The Precarious Balance*, pp. 90, 111, 194, 208, 234,
237.

agreement of 1915 was a consequence of the British occupation of Egypt in 1882; it showed, in other words, how developments outside Europe impinged on traditional European policies. When Salisbury disinterested himself in the Straits in 1895, it was because the hold Britain had acquired over the Suez canal assured the connexion with the empire overseas which, for him, was England's essential interest.[1]

When, at the close of the nineteenth century, the great movement of European expansion and encroachment in Asia and Africa reached its peak, the result, it was generally believed, would simply be to transpose the European balance of power, as it had developed during the past four centuries, from a European to a global plane. Ever since the sixteenth century, when the balance of power centred round the struggle for domination over Italy, the circle in which it operated had gradually widened as the older powers in the heart of Europe, in their efforts to maintain an equilibrium, called in new areas and new forces to counterbalance the old. Russia was brought into the European concert at the time of Peter the Great as a consequence of the strivings of the powers to prevent French hegemony; Turkey had already been mobilized against Charles V. The conflicts of England and France in the new world in the eighteenth century carried the balance of power overseas. In this way what began as a European system seemed gradually to be merging into a world system. In the nineteenth century this process appeared to be developing further. Was it not strikingly confirmed by the partition of Africa? It was not merely

1. For the question of Constantinople and the Straits, cf. A. J. P. Taylor, *The Struggle for Mastery in Europe, 1848–1918* (Oxford, 1954), pp. 359, 382, 540–3. Already in 1895 Salisbury had said he was ready to let Russia have Constantinople; according to the Committee of Imperial Defence in 1903, the opening of the Straits 'would not fundamentally alter the present strategic position in the Mediterranean'.

that the European nations each claimed a share in the spoils of Africa for fear lest their relative standing in the concert of powers should be diminished; more than that, it seemed necessary that Africa should be partitioned in order that the balance of power should continue to function as of old, but now not merely on a European but on a global plane.

This was the view that was prevalent as the twentieth century opened. Europe had simply 'overflowed' its banks into the world beyond. Although the stage of international politics now included the whole world, 'the driving forces were still the same'; all that was happening was the transformation of the balance of power in Europe into 'a balance embracing the whole world', 'the projection of the European system on to the world outside'; but no one doubted that 'the ultimate decisions would be made in Europe'.[1]

The events of the last fifty years have shown that these views were based on an illusion. They expressed a temporary constellation of forces, which had not applied a generation earlier and ceased to apply a generation later. During the middle years of the nineteenth century the prevalent attitude towards Europe's future had been distinctly pessimistic; from de Tocqueville to Constantin Frantz writer after writer had predicted the decline of Europe and the rise of Russia and the United States as the two great world-powers.[2] After 1870 pessimism gave way to optimism. There were two main reasons for this. The first was the immense advantage which appeared to

1. cf the admirable summary of German views in L. Dehio, *Germany and World Politics in the Twentieth Century* (London, 1959), pp. 42–60.

2. There has been a tendency recently to single out de Tocqueville's famous statement at the conclusion of his *Democracy in America* (1835) as exceptional. This was, however, far from being the case. The discussion of the future of Europe was continuous and lively and gave rise to an extensive literature of considerable intrinsic interest.

be accruing to the European powers from the great industrial and technological revolution which was under way. The second was the apparent restoration of the European system. The Crimean war, most people had believed, had shaken it to its foundations; but by 1871, contrary to all expectations, it appeared to have been firmly re-established. By setting up the new German empire as a solid and powerful block in the very centre of the continent, Bismarck seemed to have restored the position and given Europe a new accession of strength.

This view was not, of course, entirely wrong. As the German attempts in 1914 and 1939 to break through into the ranks of the world-powers were to show, Bismarck had created a state sufficiently strong not only to dominate Europe, but also to challenge and compete on terms of near-equality with the great extra-European powers. At the same time, it overstated Bismarck's achievement. Like Frederick II, Bismarck was adept in the art of exploiting a deadlock between the existing great powers in order to improve Prussia's position; but the Reich he created, in spite of the role it played in Europe, never really attained the same rank or magnitude as the great world-powers.[1] Moreover, there were other important facts in the situation. In the first place, the dynamic force of industrialization and technology was not, and could not be, confined to Europe. If in the twenty years after 1870 the great industrial nations of western Europe had forged ahead, the most significant feature after 1890 – apart from the beginnings of industrialization in Japan – was the accelerated tempo of industrial production in Russia and the United States. If the latter had hitherto advanced relatively slowly, between 1890 and 1914 it not only caught up with, but rapidly overtook its European rivals; while the industrial expansion of Russia – though starting off, of course, from a much lower level – showed an annual rate

1. cf. Dehio, *The Precarious Balance*, pp. 111, 212, 217.

of increase in the same period exceeding that of any other power, including the United States.[1]

Thus from about 1890, the overhauling of Europe by Russia and America, foreseen earlier but apparently checked, was resumed and intensified. It would be difficult to exaggerate the importance of this change. Although its victories in 1870 and its rapid industrialization had raised Germany to a new eminence, it was also, in view of the rising power of the United States and Russia, in a precarious position in the longer term, aware of its great potentialities but aware also that it had a definite time-limit within which to exploit its superiority; and this fact imparted an ebullient quality to German policy from the accession of William II in 1888 to the days of Hitler, which played a large part in determining the actual course of events. If Germany wanted to take a place 'in the future order of world states', Hitler argued in 1928, and not to end up as 'a second Holland or a second Switzerland', it must act quickly, for 'with the American Union a new power of such dimensions has come into being as threatens to upset the whole former power and order of rank of the states'.[2]

1. I have assembled some of the figures in the *Propyläen Weltgeschichte* (ed. Golo Mann), vol. VIII (Berlin, 1960), pp. 707–8. United States coal production was well below that of Great Britain in 1890, but by 1914 it equalled that of Great Britain and Germany combined. No less significant was the differential growth of population. Germany's population of 56 million surpassed that of every other European state, save Russia; but like that of most of western Europe its birth rate was beginning to fall before 1914, whereas the population of Russia rose from 72 to 116 million between 1870 and 1914 and that of the United States to almost 80 million. The population of the United States at the beginning of the twentieth century was still only a little more than two-thirds of that of Germany and England combined, but its industrial production surpassed their combined total.

2. cf. *Hitler's Secret Book* (New York, 1961), pp. 83, 100, 103, 158.

The rise of the United States, and parallel with it, the rise of Russia to world-power were in fact the decisive events ushering in a new period in world politics.

Already in 1883, at the very height of the age of European expansion, the eclipse of Europe had been foreseen by Sir John Seeley.[1] Russia and the United States, said Seeley, were already 'enormous political aggregations'; once their potential was mobilized by 'steam and electricity' and a network of railways, they would 'completely dwarf such European states as France and Germany and depress them into a second class'. As for England, Seeley clearly hoped that by some sort of 'federal union', transforming the colonial empire into a 'Greater Britain', it would join 'Russia and the United States in the first rank', just as Frantz before him had hoped that federation would save Europe and enable it to retain parity with the two world powers on its flanks; but Seeley was acutely aware that unless England succeeded in carrying through this transformation, it also was destined to fall back into the ranks of the 'unsafe, insignificant, second-rate'. Among Seeley's contemporaries, however, there were few who shared his vision. For this there were three main reasons. Partly, it was due to the recuperative strength of Bismarck's Germany and to the last great outburst of European expansion, which seemed to make nonsense of predictions of European decline. Partly, it was due to the tendency of Europeans to regard only European questions as of decisive importance in world affairs. And, finally, it was due to an inherent tendency to underestimate the United States as a factor in international relations, and in particular to regard it as disinterested in affairs outside the American continent.

None of these assumptions will stand closer examina-

1. cf. J. R. Seeley, *The Expansion of England* (2nd ed., London, 1919), pp. 18, 87–8, 334, 349–50.

tion. We have already seen that the effects of Bismarck's unification of Germany were limited in time. As for the United States it is true that, ever since the days of Washington, it had been a principle of American policy to steer clear of European entanglement; but that principle, though from a narrowly European point of view it may have looked like isolation, was by no means the same thing as withdrawal from affairs in other quarters of the world. Nor did it mean that the United States was not ready to make use of any complications in Europe in which its rivals in world policy became involved in order to prosecute its own interests; this it had done in 1854 at the time of the Crimean war, and was to do again in 1871, and more blatantly still in 1901, when England was entangled in the Boer war and confronted by a hostile Europe.[1] Nothing could be more misleading in this connexion than the tendency, which is still prevalent, to regard the emergence of the United States from continental isolation as a phenomenon of the recent past. Imperialist traditions and, coupled with them, a determination to play an active part in international politics, reach back to the beginnings of the history of the United States; they flowed underground for a generation after the civil war, while the United States entered a period of consolidation and began the intensive development of its internal economy; but they were not, as is often suggested, something which exploded suddenly and without precedent in 1898.

It is true that American imperialism, in its early phases, had concentrated on ensuring United States' control of the North American continent; after the Louisiana purchase Texas, Oregon, California, Cuba, Mexico, and Canada were its immediate objectives. But American policy was never exclusively continental in outlook. From the beginning it looked out across the Pacific to Asia, and the

1. For further details, cf. my article, 'Europe and the Wider World in the Nineteenth and Twentieth Centuries', in *Studies in Diplomatic History*, ed. A. O. Sarkissian (London, 1961), p. 368.

acquisition of the western and north-western seaboard, of California and Oregon, was always viewed in relation to Pacific politics and not merely as a rounding off of the continental territory. As early as 1815 Captain David Porter, who had ventured into the Pacific during the war of 1812 in search of British prizes, wrote to President Madison: 'We border on Russia, on Japan, on China. We border on islands which bear the same relation to the north-west coast as those of the West Indies bear to the Atlantic states . . .'[1] It was a theme that did not die. By 1821 the American navy had begun operating a squadron off the west coast of South America; by 1835 intercourse with China and the East Indies had reached a point which justified the establishment of a separate East India squadron. It was in the Pacific that the United States first trod the path to world power; but by mid-century its gaze extended beyond the Pacific. Americans were becoming conscious of the unity of the forces seeking expansion and their meaning in terms of a universal American empire; and nowhere was this more colourfully expressed than in an article penned by the southern journalist, J. D. B. De Bow, in 1850:

We have a destiny to perform, a 'manifest destiny' over all Mexico, over South America, over the West Indies and Canada. The Sandwich Islands are as necessary to our eastern as the isles of the gulf to our western commerce. The gates of the Chinese empire must be thrown down by the men from the Sacramento and the Oregon, and the haughty Japanese tramplers upon the cross be enlightened in the doctrines of republicanism and the ballot box. The eagle of the republic shall poise itself over the field of Waterloo, after tracing its flight among the gorges of the Himalaya or the Ural mountains, and a successor of Washington ascend the chair of universal empire![2]

This fervid, expansive imperialism, with its multiple thrusts which recognized no geographical limits, brought

1. cf. R. W. van Alstyne, *The Rising American Empire* (Oxford, 1960), p. 125.  2. ibid., p. 152.

the rising American empire into contact, and often into conflict, with the other imperialisms of the nineteenth century, French and Spanish, British and Russian. Its result was a shift in the axis of world politics. Europe did not, of course, cease forthwith to be a main centre of international rivalry, but it was not the only one, and it was rapidly ceasing to be the decisive one. The Anglo-Russian conflict in central Asia—in Persia and Afghanistan – added one new extra-European dimension; the rivalry of the powers in the Far East added another. As Constantin Frantz observed in 1859, the balance of power between the European states could now be altered 'not only on the Eider and the Po but on the Amur and in Oregon as well',[1] and it was possible to see that it was in the wider sphere, where Europe's role was relatively restricted, and not on the narrowing European stage, that the great decisions of the future would be made. But although the interrelationship between politics in different areas or regions is plainly discernible from an early date – it existed, indeed, already at the time of the Seven Years War[2] – the fusion of the different areas into one global political system only came about at the close of the nineteenth century.

In the last analysis, the switch to a global system of international politics was a result of the development of world communications – a development which, as Seeley pointed out, meant that the Atlantic ocean had 'shrunk till it seems scarcely broader than the sea between Greece and Sicily'.[3] This is the great change we see if we compare the world in 1815 and in 1900. After 1815 political events were played out on two stages, interlocked but separate; the wider stage of world politics emancipated itself from the narrower European stage to which it had long merely formed a background, and while the two great powers on

1. H. Gollwitzer, *Europabild und Europagedanke* (Munich, 1951), p. 377.
2. cf. above, p. 19.              3. Seeley, op. cit., p. 345.

the flanks of Europe, England and Russia, played their part in both theatres, the United States was still confined to the former, while the European continental powers acted wholly or predominantly in the latter.[1] And between the different theatres – this is one explanation of the generally peaceful nature of international relations for most of the nineteenth century – there was space for all. Even the Russian occupation of the immense Amur region of China in 1860, for example, did not disturb the friendly relations between the United States and Russia, for as Cassius Clay, the American ambassador in St Petersburg said, the Far East was large enough for both.[2] By 1900 this was no longer the case. The spaces between the different theatres had been filled in; the areas of the globe had shrunk; and although to the European powers it seemed as though this shrinkage, by bringing the whole world within their grasp, was putting them in a position to regulate the world's affairs in their own interests, in accordance with the principles of the European balance of power, in reality its result was to bring them face to face with powers of continental stature, which overshadowed them. Moreover, the world-powers did not accept the validity, in the spheres in which they operated, of the traditional European system. When, at the beginning of 1917, President Wilson proclaimed that 'there must be not a balance of power, but a community of power, not organized rivalries, but an organized common peace',[3] he was in effect giving notice that, in an age of world politics, the old structure of international relations had become obsolete.

1. cf. Dehio, *The Precarious Balance*, pp. 179–80.
2. cf. A. Parry, 'Cassius Clay's Glimpse into the Future', *Russian Review* 11, ii (1943), p. 54.
3. *The Public Papers of Woodrow Wilson* (ed. R. S. Baker and W. E. Dodds), *The New Democracy*, vol. 11 (New York, 1926), p. 410.

3

When did this change come about? We hear much in the closing decades of the nineteenth century of 'world policy'; but it would be a mistake to take the new catchword too literally. What we call the age of imperialism was to begin with only a new phase of European expansion, a further stage in the extension of the European balance of power into regions hitherto untouched, a last hasty effort to secure new leverage against European rivals, or to turn their flank, by reaching out into the few remaining areas which were still free from European control. The age of world politics, in the sense in which we understand the term today, was still in the future; the limiting factor was still the European equilibrium, and every political move in Asia or Africa was coolly judged in the light of the repercussions that would ensue in Europe. Thus Germany's first major intervention in the Far East, its participation in the protest of the European powers against the treaty of Shimonoseki in 1895, was undertaken with the purely European objective of weakening Russian pressure on the German eastern frontier.[1]

What brought about the decisive change was the entry on the scene of Japan and the United States between 1895 and 1905. The European powers had been able to intervene in Africa and partition it in conformity with their own ideas of the balance of power because neither Russia nor the United States was directly involved in African political affairs. When, after 1895, they turned to China and began the process of carving it up on the African model, they quickly found that they were confronted by an entirely different political situation. In the Far East it was not only the European powers that called the tune.

1. cf. Taylor, op. cit., p. 357; Oncken, *Das deutsche Reich und die Vorgeschichte des Weltkrieges*, vol. II, p. 431; A. S. Jerussalimski, *Die Aussenpolitik und die Diplomatie des deutschen Imperialismus* (Berlin, 1954), p. 486.

The three countries immediately concerned were Japan, Russia, and the United States. They were directly concerned because they were all Pacific powers; for here the position of Russia, with a contiguous land-frontier and a coastline on the Pacific as long as that of Norway, was basically different from that of the other European powers. In Asia Russia acted in the capacity of an Asian power, for in Asia – as the well-known publicist, Michael Katkov, took pains to emphasize – the Russians were not 'foreign intruders from afar, as England is in India', but were 'as much at home as in Moscow'.[1]

If China and the Far East had developed, like Africa, into a sort of dependency partitioned among the European powers, it is unlikely that their fate would have had much effect on the existing international equilibrium. What prevented this happening was the reaction of the non-European powers, Japan and the United States, which were not prepared to stand aside while the European powers disposed of areas which they regarded as vital to their own prosperity and security. Thus the events in the Far East between 1898 and 1905 proved to be a turning-point. The threat of the partition of China, the fear of the mainland falling under European control, spurred the extra-European powers into action. The result was the emergence of a system of world politics which ultimately displaced the European system. This was the significance in terms of world history of the events in Asia in these years.[2] Never before had European, American, and Asiatic policies interacted in the same way. The world received in 1905 a first glimpse of the future global age.

What was new in the situation which developed in the Far East between 1898 and 1905? It was not, as is so often said, that the United States emerged from isolation, for, as has already been noted, the United States had been inter-

1. cf. E. Hölzle, *Geschichte der zweigeteilten Welt*, p. 125.
2. cf. H. Holborn, *The Political Collapse of Europe* (New York, 1951), p. 69.

vening actively in Far Eastern affairs, mostly as the rival of Great Britain, for two or three generations. It was not, either, that it resulted in the first war between a western and a non-western great power, though Japan's rise to great-power status and its victory over Russia were certainly important events in the history of the relations of Asia and Europe. Nor was it simply, as has recently been asserted,[1] that 'the search for power and equilibrium' was 'extending beyond the narrow circle of European powers'. This formula is perhaps adequate to describe the situation from 1882 to 1895 or 1898; but it is unsatisfactory as a description of the situation after 1898 because it starts from the assumption, which was rapidly losing validity, that Europe was still the centre, whereas in reality the primacy of Europe was drawing to a close, its sphere of action was contracting as new extra-European powers came to the fore, and the European system of balance was ceasing to determine the structure of world politics. To gauge the importance of the Far Eastern crisis we must look further afield; above all, we must see it not merely in terms of European politics – though its repercussions on European alignments are, of course, a matter of common knowledge – but in the broader perspective of world history.

From this point of view – and leaving out, for the present, the stimulus they gave to Chinese nationalism[2] – it may be said that the events in the Far East between 1898 and 1905 had five important consequences. First, they marked the end of the long friendship and understanding between Russia and the United States, and brought them face to face as rivals in the Pacific. Secondly, they finally established the Far East as a centre of international rivalry and conflict which, though European powers might regard it as secondary and subordinate, was for the extra-

1. F. H. Hinsley, *Power and the Pursuit of Peace* (Cambridge, 1963), p. 257.
2. cf. below, pp. 154–5, 160, 163–4.

European powers, particularly the United States, in many respects more important than Europe itself. Thirdly, they saw the formation of a permanent link between European affairs and world affairs and, over a longer term, the gradual subordination of the former to the latter. Hence they implied, fourthly, that Europe was losing its primacy; the world upon which it had pressed for a century now began to press in on Europe, until finally Europe, which had tried to make the world its appendage, became the appendage of the two world-powers, the United States and the Soviet Union. And lastly, they were a turning-point in the process by which the system of balance of power, European in origin and dependent for its continuance upon the pre-eminence of Europe, gave way to the system of world polarity, division among a multiplicity of competing and self-balancing interests to the establishment of great self-contained, continent-wide power blocks, from which rigid iron curtains excluded all extraneous powers. At the end of this development, and symbolizing the change, stand the Berlin wall of 1961 and the United States' action to enforce the withdrawal of Russian rocket bases from Cuba in 1962.

By 1905 we can perceive fundamental changes in the world situation. The driving forces were no longer the same as previously; the ultimate decisions were no longer made in Europe. This was the most significant result of the Russo-Japanese war. When in 1902 the English government allied with Japan it looked as though it had pulled off a clever manoeuvre against Russia, but in reality it had called in a force it could not control. Churchill might confidently predict that Japan would remain dependent on Britain for many decades to come;[1] but on any longer-term view it was Japan which henceforward made use of its British alliance to promote its own interests – as it was

1. In his speech in the House of Commons on 17 March 1914; cf. Keith, *Speeches and Documents on Colonial Policy*, vol. 11, pp. 351-2.

to do even more effectively after the outbreak of war in Europe in 1914 – and not vice versa. Hitherto the protagonists in the Far East had been England and Russia, and the latter had usually had the sympathy and benevolent neutrality, if not the active support, of the United States. After 1905 the three powers directly concerned in the Far Eastern conflict were United States, Russia, and Japan, and the European powers were gradually edged out.

The United States had become a Far Eastern power in 1898 when it annexed the Philippines and Guam, and it was the United States which, in the following year, warned the powers to keep their hands off China. Hay's 'open door' note of 6 September 1899, which enunciated the principle of the integrity and inviolability of China, is noteworthy as the first occasion on which the United States made a pronouncement of a general character concerning affairs outside the American continent.[1] It is significant that it was followed up a few years later, at the time of the Morocco crisis, by the first American intervention in European affairs.[2] Nevertheless it was in the Far East, where its interests were directly involved, that the United States began to assume the role of a world-power, and it was its Far Eastern interests that brought it into conflict with Russia.[3] Hay's note of 1899, though it served notice generally that the United States was not prepared to see eastern Asia turned into a battleground of European

1. There has been much discussion of the origins of the 'open door' policy; cf. particularly A. W. Griswold, *The Far Eastern Policy of the United States* (New York, 1938), pp. 36–77; P. Varg, *Open Door Diplomat. The Life of William W. Rockhill* (Urbana, 1952); C. S. Campbell, *Anglo-American Understanding, 1898–1903* (Baltimore, 1957), pp. 151–79. I am inclined to agree with F. R. Dulles, *America's Rise to World Power* (New York, 1954), that the view that the open door policy was British-inspired is not borne out by the facts.

2. cf. Hölzle, *Archiv für Kulturgeschichte*, vol. XL, p. 354.

3. cf. E. H. Zabriskie, *American-Russian Rivalry in the Far East, 1895–1914* (Philadelphia, 1946).

power politics, was directed primarily against Russia, the only power at that date which could effectively challenge the United States in the Pacific region.

The rivalry in east Asia between Russia and the United States was not yet direct and open; it took the form, rather, of tacit American support for Japan and an Anglo-American *rapprochement,* and it was modified and mitigated when the unexpected Japanese victory in 1905 demonstrated that Russia was not the only power capable of threatening American interests in the area. Nevertheless it marked a diplomatic revolution of first magnitude. For a hundred years the two powers had supported each other against England; now, as England's power passed its zenith, they came face to face across the Pacific. Thus began a conflict of interests which was eventually to spread to Europe, to south-east Asia and to the Middle East, until in the end it divided the world into two hostile camps. What today we too easily simplify as an ideological conflict – the so-called 'cold war' – had its origins in the new power constellation which began to take shape at the beginning of the twentieth century.

4

If it is important to be aware of the threads leading from 1898 to the final crystallization of Russo-American rivalry in 1947, it is also important not to exaggerate their immediate effect. It is only when we look back in retrospect that the wider significance of the events of 1898–1905 stands out clearly. Though the period after 1898 was the beginning of the post-European age, it was also, as was pointed out earlier,[1] the end of the European age, and the European powers did not surrender their inherited pre-eminence without a struggle. Thus the first half of the twentieth century, in international relations, is a period of the utmost confusion, in which a new system is strug-

1. cf. above, p. 31.

gling to be born and the old system fighting hard for its life.

There can be no question, in the present context, of considering in detail the long and intricate history of this confused period. For the most part historians have tended to emphasize the second or European aspect – namely, the German bid to reorganize Europe as a continental empire, capable of holding its own against the vast American and Russian trans-continental empires – and in this, it may fairly be said, they have only reflected what was undoubtedly the main preoccupation of the European powers, including England and Russia, at the time. It is nevertheless a one-sided view. Europe was only one element in an international system of far greater complexity, and the great extra-European powers, with their world-wide interests, inevitably saw the international situation between 1905 and 1914 in a different light from the European powers, particularly as they had never been part of the European system. From the point of view of the United States – and, indeed, of Japan – the German bid for hegemony in Europe was certainly not the central issue. For Americans the tangible threat of British sea-power was more real than the hypothetical threat of German land-power; and though, even before 1914, there was a small minority in the United States which argued that, if Germany got its way in Europe, it would sooner or later challenge the United States in the western hemisphere, from an American point of view the question of the European balance of power remained, right down to 1917, a secondary and local issue, which did not vitally affect American interests.[1] More significant, for the United States,

1. The question, involving as it does the causes of the American entry into the First World War, is too big to discuss here; but I think the statement in the text is a fair representation of the conclusions of most American historians. It seems to me important to make this clear in view of the recent tendency to treat Anglo-American understanding as a basic element in United States policy from about 1902, and to exaggerate – in the light of the 1941 situation – United States'

than the balance of power in Europe was the balance of power in the western Pacific, or rather the threat, which Americans began to discern after 1907, of the establishment of Japanese hegemony. What was happening – if we look at the situation between 1905 and 1917 from a global rather than from a narrowly European point of view – was not merely an attempt to impose a new statute on Europe but also a bold attempt to reshape the balance in the Far East; and it was the juxtaposition of these issues, and their interconnexion, that was the distinctive feature of the situation.

The role which Germany and Japan began to play in the early years of the twentieth century was a function of the new world situation. The rise of the new Germany, the very success of Bismarck in launching the new Reich in 1871, was itself a consequence of the rivalry of Russia and Great Britain outside Europe; against their united opposition Bismarck's stroke would scarcely have been possible.[1] On the other hand, the rise of powerful new national states in Europe – namely, united Germany and united Italy – affected the ability of both Russia and England to maintain their position in the wider world. It was by exploiting Russian and English preoccupations with the dangers by which they believed they were threatened in Europe that Japan consolidated its position in Asia, just as the United States took advantage of the South African war, Britain's conflict with France at Fashoda and the Anglo-German naval rivalry to enunciate a more ambitious Monroe doctrine, eliminate British influence from the Panama canal zone and round off its hold over the Caribbean. What was developing, in short, was a

preoccupation with the German 'menace'. This is reading history backwards; the position of the United States, in relation to Great Britain and Germany, and in relation to European questions as a whole, was far less clear-cut.

1. cf. W. E. Mosse, *The European Powers and the German Question, 1848–1871* (Cambridge, 1958), pp. 372, 374.

situation of which Seeley had been half aware in 1883, but from which he failed to draw the right conclusion. Seeking to account for the success of England in its conflicts with Spain, Portugal, Holland, and France for 'possession of the New World', Seeley came to the conclusion that the explanation lay in the fact that England alone had not been 'deeply involved in the struggles of Europe'; 'out of the five states which competed for the new world', he maintained, success fell 'to that one which was least hampered by the Old World'.[1] It was an acute observation; but it is another question whether England was an exception to the rule. In 1866 Disraeli had proudly proclaimed that England had grown out of the European continent; it was more an Asiatic than a European power.[2] But after 1898 Disraeli's boast was no longer true. The rising power of Germany drew Britain back to Europe, just as Russian policy after its reverses in the war with Japan, swung back to Europe. 'We must put our interests in Asia on a reasonable footing,' Izvolsky remarked in 1907, 'otherwise we shall simply become an Asiatic state, which would be the greatest disaster for Russia.'[3]

What emerges clearly is that both England and Russia were hampered in their European policy by the need to consider their interests outside Europe, just as their ability to defend their interests in the wider world was affected by their need to look over their shoulders at Europe. It is true that Russia was affected less seriously by this ambivalence

1. Seeley, op. cit., pp. 108–13.
2. cf. W. F. Monypenny and G. E. Buckle, *The Life of Benjamin Disraeli*, vol. 11 (2nd ed., London, 1929), p. 201.
3. cf. B. H. Sumner, 'Tsardom and Imperialism in the Far East and the Middle East, 1880–1914', *Proceedings of the British Academy*, 1941, p. 64. Whether Izvolsky's judgement was correct, is another matter. One of the best of living historians of Russia has suggested, on the contrary, that, having regard to the prospects opening out before it in Asia, Russian policy between 1907 and 1914 was 'too European'; cf. R. Wittram, 'Das russische Imperium und sein Gestaltwandle', *Historische Zeitschrift* CLXXXVII (1959), p. 591.

than England, for Russia's established position in central Asia, which no country – not even Japan – was in a position to challenge, provided a firm foundation for its world power, whereas England's world position depended on a preponderance at sea which was no longer absolute, and on a thinly spread overlordship over recalcitrant peoples, among whom it was not difficult for hostile powers to stir up disaffection. Even so, Seeley's observation, as a generalization, retains its value. It is true that Germany and Japan were able to forge ahead, in part at least, because they were not diverted by interests outside their respective spheres; it is true that its ability to dissociate itself from European complications greatly helped the United States, particularly in relation to Britain, to concentrate on its world interests; and, finally, it is true that the decline in British power was a reflection of the inability of England, after the 1890s, to find a solution to the problem which had faced it ever since the eighteenth century, of striking a balance between its world interests and its European interests.

As circumstances and its own accumulating strength enabled Germany to play an increasingly important part in international relations, it was natural that European affairs should again loom large; for although German diplomacy under Bülow and Bethmann Hollweg sailed under the banner of 'world policy', the incidents in which Germany was involved overseas – Tangier, the first Morocco crisis, Agadir, the Baghdad railway – were undertaken in all essentials for the purpose of giving the German government increased leverage in Europe, and Europe always remained the focal point of German policy. Consequently historians have usually agreed that after 1905 international politics once again relapsed into their old forms.[1] In particular, they have pointed out that the war that broke out in 1914 was at the beginning not a world war but a great European war, and that it only turned into

1. Holborn, op. cit., p. 70.

a world war in 1917. Thus, even as late as 1914, it is argued, European questions retained their primacy; what led to the war was neither European imperialism, nor the expansion of the economic and colonial interests of the European powers, nor the involvement of the great extra-European powers in their rivalries, but – by this time an anachronism – their age-old European antagonisms.[1]

Nevertheless this view, though perhaps true in a literal sense, leaves some important factors out of account. The war of 1914 is commonly regarded as one of a long series of struggles, beginning with Charles V and Philip II of Spain and ending with Hitler, to establish the hegemony of one power over Europe.[2] But whatever may have been true of earlier struggles for hegemony, this is only a partial explanation of German war aims in 1914. That Germany sought to establish its dominion over Europe is not in doubt; but its purpose – unlike that of earlier contenders for European hegemony – was not European, but, as Plehn formulated it in 1913, to 'win our freedom to participate in world politics'.[3] The difference is important, for it indicates that the war, though fought out for the most part in Europe, was from the beginning conceived of as a world war. The reason, clearly enough, was that the displacement of the older structure of power by a new one provoked a new response, the rise of the great world powers created a new challenge.

Thus we may say that the First World War was Germany's response to a new constellation of world forces – as, indeed, was Hitler's war a quarter of a century later. Germany's war aims, spelled out in detail as early as 9 September 1914, were the creation of two vast empires, the

1. E. Kessel, 'Vom Imperialismus des europäischen Staatensystems zum Dualismus der Weltmächte', *Archiv für Kulturgeschichte* XLII (1960), p. 243.

2. This is the underlying thesis of Dehio's book, *The Precarious Balance*; cf. for example p. 263.

3. Plehn's statement is cited by Dehio, *Germany and World Politics in the Twentieth Century*, p. 16.

one in the heart of Europe, the other in central Africa.[1] The realization of these aims, as Tirpitz and the builders of the German navy were well aware, was bound to bring Germany up against England; and the English reaction, in seeking to contain Germany, as a century earlier it had sought to contain Napoleon, by a continental coalition, forced the conflict into the classical mould of a struggle for European hegemony. But though these facts might make it necessary to trim down England in the process, it is probably true that Germany's aim was not to destroy England, as Napoleon's had been, but to ensure its own entry into the 'future concert of world-powers' by setting up a German empire to equal the British empire and the developing world empires of Russia and the United States. Moreover, from the beginning Germany conducted the war as a world war. Already on 2 August 1914, before the outbreak of hostilities, its plans were laid: intervention in India, Egypt and Persia, support for Japan and the promise of an exclusive Japanese sphere of interest in the Far East, insurrection in South Africa, even a scheme to win over the United States by the prospect of annexing Canada.[2]

Although the war of 1914 started in Europe, it was thus from the beginning a world war in conception and in plan. Moreover, it is easy to underestimate its effects in Asia. It was not only that Japan, which declared war as early as 23 August 1914, immediately seized the German concessions in China and the islands belonging to Germany in the northern Pacific; it also exploited to the full the fact that Russia and Britain were tied down in Europe to confront China with the notorious 'twenty-one demands'. It would take us too far to consider in detail the diplomatic manoeuvres pursued by Japan—its secret negotiations with

1. The full details of the German plans were brought together from the official German documents and published by Fritz Fischer in his important book, *Griff nach der Weltmacht. Die Kriegszielpolitik des kaiserlichen Deutschland, 1914–1918* (Düsseldorf, 1961), pp. 107–12.

2. ibid., pp. 93–4.

Germany, for example, the project of a German–Russian–Japanese *bloc*, and the Russo-Japanese alliance of 1916 as a guarantee against United States interference.[1] What is certain is that the effects of the war on the power situation in the Far East – particularly when the Russian revolution in 1917 gave Japan further possibilities of building up its ascendancy – were no less revolutionary than those in Europe. By 1918, even before the end of the European war, Wilson was already girding himself to challenge in earnest the expansion of Japan.

In Europe the effect of the war was to destroy for all time the basis of the European balance of power. In 1815, after the Napoleonic wars, an attempt had been made to ensure stability by constructing a new equilibrium; in 1919 that was not the case. At the peace conference in Paris there was no question of restoring a self-contained European system, such as had existed before the war; probably it was no longer practicable.[2] It is true that the absence of Russia, in consequence of the 1917 revolution, and of the United States, in consequence of the American withdrawal into isolation after the fall of Wilson, created the illusion that the European balance of power still existed; but only the dismemberment of Germany could have achieved a real equilibrium, and already the incipient world conflict, the western powers' fear of communist infiltration, precluded dismemberment.[3] On the other

1. cf. A. Whitney Griswold, *The Far Eastern Policy of the United States* (New York, 1938), pp. 176–222; J. M. Shukow, *Die internationalen Beziehungen im fernen Osten, 1870–1945* (Berlin, 1955), pp. 173–95; E. Hölzle, 'Deutschland und die Wegscheide des ersten Weltkrieges', *Geschichtliche Kräfte und Entscheidungen*, ed. M. Göhring and A. Scharff (Wiesbaden, 1954), pp. 266–85. These works refer to most of the specialized literature. I have been unable to see O. Becker, *Der ferne Osten und das Schicksal Europas, 1907–1918* (Leipzig, 1940).

2. cf. H. Holborn, 'Die amerikanische Aussenpolitik und das Problem der europäischen Einigung', *Europa: Erbe und Aufgabe*, ed. M. Göhring (Wiesbaden, 1956), p. 303.

3. cf. L. Kochan, *The Struggle for Germany, 1914–1945* (Edinburgh, 1963), pp. 5–6, 9, 12.

hand, the *cordon sanitaire* round Germany, constructed and directed by France, was incapable of performing the function for which it was designed, as the meteoric rise of Hitler was soon to show. By 1918, in short, the power of the European nations had withered, and the decisive role had passed to the two great extra-European powers on their flank. The total defeat of Hohenzollern Germany, precluding a negotiated compromise peace, was the result of the crushing superiority of the United States, just as in 1945 the total defeat of Nazi Germany was due to the United States and Soviet Russia. Europe alone, even if Great Britain is included in Europe, was no longer able to solve its own problems.

5

It is therefore no exaggeration to say that the entry of the United States into the war in 1917 was a turning-point in history; it marked the decisive stage in the transition from the European age to the age of world politics. It is a turning-point in another way also. After the Bolshevik revolution in Russia in November 1917 the division of the world into two great rival power blocks, inspired by apparently irreconcilable ideologies, took tangible shape. Though it was something like two years before President Wilson identified himself with the anti-Bolshevik crusaders in the west, he and Lenin were aware from the start that they were competing for the suffrage of mankind, and it was to prevent Lenin gaining a monopoly of the blueprints for the post-war world that, in January 1918, Wilson issued his famous Fourteen Points. 'Either Wilson or Lenin,' wrote the French socialist, Albert Thomas; 'either Democracy or Bolshevism. . . . A choice must be made.'[1] But, in spite of their rivalry, Wilson and Lenin had one thing in common: their rejection of the existing inter-

1. cf. B. W. Schaper, *Albert Thomas* (Leiden, 1953), pp. 175–6.

national system. Both rejected secret diplomacy, annexations, trade discrimination; both cut loose from the balance of power; both denounced the 'dead hand of the past'. They were 'the champion revolutionists of the age', 'the prophets of a new international order'.[1]

Here again was a decisive break. Although at a later date both the Soviet Union and the United States were to revert in practice to the old methods of power politics, by then the revolutionary principles enunciated by Wilson and Lenin had done their work. From the beginning of 1917 the conflict among the European powers was transformed from a war of limited objectives into a world-wide revolutionary and ideological struggle. The war aims of England, France, Czarist Russia, and Italy, as formulated in the secret treaties, had assumed that the war would lead to the re-establishment of a European balance of power without markedly disturbing the domestic *status quo* in any of the major belligerent nations. After the Russian revolution and the entry of the United States into the war this assumption ceased to hold good. Lenin and Trotsky, counting on universal revolution, refused to accept the permanence of a system of independent self-balancing states; Wilson had no faith in the mechanism, traditional in Europe since the defeat of the French revolution, through which the powers conducted their affairs by adjusting the claims of sovereign states against each other as they arose; and all three planned to end the balance of power, not to restore it. Thus the continuity in mood, procedures, and objectives of nineteenth-century diplomacy was irretrievably disrupted. One of the most momentous developments of the war was the simultaneous emergence of Washington and Petrograd as two rival centres of power, both of which abandoned the old

1. Harold D. Lasswell, *Propaganda Technique in the World War* (London, 1927), p. 216; R. W. van Alstyne, 'Woodrow Wilson and the Idea of the Nation State', *International Affairs* XXXVII (1961), p. 307.

diplomacy and its ruling concept, the balance of power.[1]

The character of this revolution has often been mis-judged. Historians have for the most part attributed Wilson's rejection of the concept of balance of power to an utopian moralism, for which he has been both praised and condemned. In reality, as an acute observer has remarked, it was due 'less to any alteration in ethical standards than to a shift in the centre of power'.[2] Wilson and Lenin were no less realists than Clemenceau, Sonnino, or Lloyd George; but the realities they had to deal with were different. From the point of view of the United States, the policy of territorial adjustments, annexations and com-pensations on which European statesmen relied, was of little consequence in the sense that it would neither increase American security nor improve its strategic posi-tion; and Wilson was right when he perceived, as Lenin also perceived, that the new diplomacy – the diplomacy of appeals to the people over the heads of politicians – would serve his purpose better in a rapidly changing world. The entry of the United States into the war meant, therefore, not – as allied statesmen deceived themselves into thinking –simply the mobilization at the crucial moment of another all-powerful belligerent, the addition of a decisive counter on the existing political chessboard; on the contrary, it meant the appearance on the scene of a power which had, for historical reasons, scant interest in the old European political system, which was not prepared to underwrite the European balance of power, and which had almost irresis-tible means, in the state of exhaustion to which the European powers on both sides had been reduced, of enforcing its point of view.

When revolutionary Russia, under Lenin and Trotsky, embarked on a parallel course, the breach with the past

1. cf. A. J. Mayer, *Political Origins of the New Diplomacy, 1917–1918* (New Haven, 1959), pp. 22, 33, 34, 290, where the revolu-tionary impact of Wilson and Lenin is discussed in detail.

2. Harold Nicolson, *Diplomacy* (London, 1939), p. 60.

became irrevocable. The Bolsheviks also repudiated the old system of balance of power. So far as the security of Russia was at stake, they sought it not, as Stalin was to do in 1939, 1944 and 1945, by piecemeal territorial annexations – the Baltic states, East Prussia, Bukovina, etc., – but through world revolution. By 1915 Lenin was convinced that the power of Europe was on the wane.[1] The war, he believed, would give a decisive impetus not only to the ripening of a revolutionary crisis in Europe, but also to the development of extra-European centres of power and to a colonial awakening which would decisively weaken the European countries that had hitherto been dominant. Like Wilson, in short, but from a very different starting-point, Lenin moved under the impact of war – a world war which involved India, China, Japan, the Arab world, and the United States – from a European to a global view of international politics, which he set out to embody in Bolshevik doctrine and strategy. In many ways the most significant feature both of Wilson's programme and of Lenin's is that they were not European-centred but world-embracing: that is to say, both set out to appeal to all peoples of the world, irrespective of race and colour. Both implied a negation of the preceding European system, whether it was confined to Europe or whether it spread (as it had done during the preceding generation) over the whole world. And both quickly fell into bitter competition. Lenin's summons to world revolution called forth, as a deliberate counter-stroke, Wilson's Fourteen Points, the solidarity of the proletariat and the revolt against imperialism were matched by self-determination and the century of the common man. These were the slogans under which a new international system, different in all its basic tenets, took over from the old, and which preclude the view, still occasionally propounded, that it was merely 'the same system of states in a new phase of its development'.[2]

1. cf. Mayer, op. cit., pp. 298–300.    2. Hinsley, op. cit., p. 357.

In this way, to the divergence of political interests which had begun to affect Russo-American relations at the close of the nineteenth century was added a deep ideological cleavage, and each side set up a banner around which to gather its forces. The reversal that followed – the withdrawal of the United States into isolation, the weakening of Soviet Russia under the stress of civil war – does not detract from the importance of this turning-point. It allowed Japan under Tojo to make, and Germany under Hitler to renew, its bid for a place among the world-powers; but the result was to establish the primacy of the Soviet Union and the United States even more definitely than before.

After 1945 the division of the world between Russia and America went on apace. It would doubtless be a mistake to regard the resulting cleavage of the world into two conflicting power blocks as final; but in spite of a neutralist group attached to neither side, and in spite of differences between Russia and China, on the one side, and between the United States and its associates in western Europe, on the other, it is the situation with which we are confronted today. Much has been written in recent years of the dissolution of the monolithic blocks which dominated the world for a decade after 1947. The fact remains that the Soviet Union and the United States are so far ahead of the rest of the world in nuclear weapons, and the resources to construct nuclear weapons, that no other country is in a position to challenge their preponderance within a foreseeable future. Even if 'bipolarity', which stamped its imprint on the period after the Second World War, may now be passing, it is evident that any new multipower system will be fundamentally different, in its structure and in its modalities, from the 'classical' one. It will not restore the old 'multinational' system with its graduated scale of comparable 'powers', and any new multipower system thus arising will have aspects very different from what those who talk in terms of a "return" to a five-power or six-power system are envisaging'.[1]

It can be said also, with no less assurance, that the system of bipolarity under which we live today is not simply a result of the conditions created by the Second World War. The gradual emergence of Russia and the United States as super-powers and the declining importance of the European states are developments reaching back before the beginning of the present century; they are one of the clearest signs of the onset of a new age. When in 1898 Germany embarked on its new naval programme, a majority on both sides believed that what was at stake was whether the balance of empire would lie with Germany or with Britain. In fact, neither was to come to pass. Little though it was realized at the time, the European powers, in reaching out to Asia, Africa, and the New World, had called on to the scene forces which were to overshadow them. The eclipse of Europe, the rise of the great extra-European super-powers, the end of the system of balance – a 'peculiar mechanism of European history . . . unmatched anywhere else in the world'[2] – and the consolidation of great, stabilized continental *blocs* in a world where the areas of free manoeuvre have disappeared and power positions have congealed: all these things, which have become so familiar in the last fifteen years, were implicit, if not yet clearly visible, in the new world situation which took shape at the close of the nineteenth century.

1. cf. J. H. Herz, *International Politics in the Atomic Age* (New York, 1959), pp. 34–5; cf. ibid., pp. 153, 156–8. This is still the best discussion of the question in its contemporary connotation.
2. Dehio, *The Precarious Balance*, p. 268.

# V

# FROM INDIVIDUALISM TO MASS DEMOCRACY

*Political Organization in Technological Society*

IN a famous 'diagnosis of our time', published in 1930, the Spanish philosopher, Ortega y Gasset, proclaimed that 'the most important fact' of the contemporary epoch was the rise of the masses.[1] It is not necessary to adopt Ortega's interpretation of the significance of this fact to share his belief in its importance. We need only look around us to see how radically the advent of a mass society has changed the context not only of our individual lives but also of the political system by which our society is organized. Here again, the closing decades of the nineteenth century, or more widely perhaps the years between 1870 and 1914, stand out as a watershed, dividing one historical period from another. As new large-scale industrial processes were introduced and new forms of industrial organization arose, necessitating the concentration of population in sprawling congested areas of smoky factories and dingy streets, the whole character of the social structure changed. In the new conurbations a vast, impersonal, malleable mass society came into existence, and the scene was set for the displacement of the prevalent bourgeois social and political systems, and the liberal philosophy they upheld, by new forms of social and political organization.

Similar conditions had, of course, already existed for some generations in a few of the areas of early industrialization – in Manchester, for example, or Glasgow, or Sheffield – but even in England they had been exceptional.

1. J. Ortega y Gasset, *La Rebelión de las Masas* (Madrid, 1930; reprinted in *Obras*, vol. VI, Madrid, 1946); Engl. trans., *The Revolt of the Masses* (1932, paperback ed. 1961).

Now the exceptional became normal, producing immediately a series of fundamental problems with which the existing machinery of government was unable to cope. Questions of sanitation and public health, for example, suddenly became urgent – how otherwise could epidemics from the slums be prevented from spreading and slaughtering thousands and tens of thousands without respect for rank or person? – and governments were compelled to take action and to construct new machinery which made effective action possible. The result was that a new philosophy of state intervention was born.[1] In Germany, Bismarck's social legislation of 1883–9 marked the turning-point. In England, the radical programme sponsored by Chamberlain in 1880 sounded 'the death knell of the *laissez-faire* system', Gladstone's cabinet of 1880–5 was 'the bridge between two political worlds'.[2] Government in its modern sense of regulation, state control, compulsion on individuals for social ends and ultimately planning, involving the development of an elaborate machinery of administration and enforcement, was a necessary outcome of the new industrial society; it had existed hardly anywhere before 1870, because it was a response to conditions which only reached full-scale development after that date.

1

It was inevitable that, sooner or later, the effects of these changes should make themselves felt over the whole range of political life and political organization. Once the state ceased to be regarded as a night-watchman whose activities should be restricted to a minimum in the interests of

1. The classical account of the change, so far as England was concerned, is to be found in A. V. Dicey, *Lectures on the Relation between Law and Public Opinion in England during the Nineteenth Century* (London, 1905).

2. cf. K. B. Smellie, *A Hundred Years of English Government* (London, 1937), p. 212.

individual freedom, once it was given positive and active instead of merely supervisory and repressive functions, once the scope of politics was widened until it embraced, in principle at least, the whole of human existence, it was only a matter of time before the machinery by which governments were elected, controlled and vested with power, was adapted to the new circumstances. Just as the resources at the command of governments in the half-century after 1815 were inadequate for solving the problems with which industrialization confronted them, so the political machinery existing down to the time of the Second Reform Bill in England, or the introduction of universal manhood suffrage in the North German Confederation in 1867, was not of a kind by which the forces of a mass democracy could be mobilized and turned to effective use.

In the first place, the conditions for which the existing political machinery had been devised were entirely different. Hitherto, as Sir James Graham pointed out in 1859, representation had been based on 'property and intelligence'.[1] In England and Wales the Great Reform Bill of 1832 had added only some 217,000 voters to the existing electorate of 435,000; and though the rising population and wealth of the country brought a further increase of about 400,000 by the time of the Second Reform Bill of 1867, even then the electorate was not much more than one in thirty in the United Kingdom as a whole. This meant not only that five out of six adult males, and by far the greater part of the working class, were voteless, but also that it was still easy – particularly before the introduction of the secret ballot in 1872 – to manipulate elections by influence, bribery, and intimidation.[2] In France conditions under Louis Napoleon were exceptional, but the disproportion here had previously been even greater.

1. Smellie, op. cit., p. 45.
2. There is an entertaining account of the 1868 election in Lynn – 'the first parliamentary election I remember' – in G. G. Coulton, *Four-score Years* (Cambridge, 1943), pp. 22-4.

Under the electoral law in force in France from 1831 to 1848, the electorate was confined to some two hundred thousand out of a population of approximately thirty million.[1] And in the relatively limited number of German states – Baden, Hesse and Württemberg, for example – where representative institutions, often modelled on those of the French charter of 1814, were permitted to function, the position was essentially the same. Nineteenth-century liberal democracy, in short, was everywhere constructed on the basis of a restricted property franchise; like Athenian democracy in the ancient world, it was really an 'egalitarian oligarchy', in which 'a ruling class of citizens shared the rights and spoils of political control'.[2]

This situation was radically altered by the extension of the franchise. Both in the German empire and in the new French republic universal manhood suffrage was an accomplished fact from 1871. Switzerland, Spain, Belgium, the Netherlands, and Norway followed suit in 1874, 1890, 1893, 1896, and 1898 respectively. In Italy, where a very limited increase in the franchise had been granted in 1882, most of the male population received a vote by a law passed in 1912; in Great Britain the same result was achieved by the Third Reform Bill of 1884, although there the principle of universal manhood suffrage had to wait until 1918 for recognition and full suffrage was not extended to women until 1928. In the areas of European colonization overseas the extension of the franchise tended, not surprisingly, to occur a good deal sooner. This was the case in New Zealand, Australia, Canada, and, of course, in the United States of America, where universal manhood suffrage was introduced almost everywhere between 1820 and 1840 with immediate effects on the political machinery. After 1869, moreover, the suffrage was gradually

1. The number varied from 166,000 in 1831 to 247,000 in 1847; before 1831 it had not reached 100,000; cf. P. Bastid, *Les institutions de la monarchie parlementaire française* (Paris, 1954), pp. 225, 227–8.
2. R. M. MacIver, *The Modern State* (London, 1932), p. 352.

extended to women, until in 1920 an amendment to the constitution enfranchised women in all states of the Union. In New Zealand, where manhood suffrage was established in 1879, women were given the vote in 1893.[1]

The effect of these changes, stated shortly, was to make unworkable the old system of parliamentary democracy that had developed in Europe out of the 'estates' of late medieval and early modern times, and to inaugurate a series of structural innovations which resulted in a short space of time in the displacement of the liberal, individualist representative system by a new form of democracy: the party state. A number of factors have combined to conceal the revolutionary nature of this transformation. The first is terminological. In England, in particular, the mere fact that the history of political parties, and the term 'party' itself, reach back in apparent continuity into the seventeenth century, has been sufficient to create the illusion that all that occurred was a process of adaptation which broadened the foundation but left the essence of the old structure standing. In the second place, current ideological conflicts have obscured the issue. In the United States and in western Europe, people have been so concerned to demonstrate that western democratic practice is the only reliable safeguard of the individual's rights and liberties, by comparison with the one-party system prevalent in fascist and communist countries, that it has seemed almost treasonable to inquire how far *all* the modern forms of government mark a breach with the representative democracy of a century ago. In this respect the currently popular distinction between liberal and totalitarian

1. For details, cf. James Bryce, *Modern Democracies*, vol. II (London, 1921), pp. 50, 188, 199, 295, 339. In England, women had been partially enfranchised in 1918; in Switzerland, they still do not have the vote. In the U.S.A. 'nearly all the southern states' (in Lord Bryce's words) 'passed enactments which, without directly contravening the constitutional amendment of 1870 designed to enfranchise all the coloured population, have succeeded in practically excluding from the franchise the large majority of that population'.

democracy is not altogether satisfactory, since, whatever its value in terms of political theory, it fails to take account of the fact that communism, fascism, and the modern western multiparty system are all different responses to the breakdown of nineteenth-century liberal democracy under the pressure of mass society.

To say this is in no way to disparage the multiparty system or to dismiss it as a mere parody of 'true' democracy (an abstraction which has never existed), but simply to point out that it must be classified and justified on its own terms and not by nineteenth-century political standards. To speak of the defence of democracy as if we were defending something we had possessed for generations, or even for centuries, is wide of the mark. The type of democracy prevalent today in western Europe – what we summarily call 'mass democracy' – is a new type of democracy, created for the most part in the last sixty or seventy years and different in essential points from the liberal democracy of the nineteenth century. It is new because the politically active elements today no longer consist of a relatively small body of equals, all economically secure and sharing the same social background, but are drawn from a vast amorphous society, comprising all levels of wealth and education, for the most part fully occupied with the business of earning a daily living, who can only be mobilized for political action by the highly integrated political machines we call 'parties'. In some cases – for example, in the 'people's democracies' of eastern Europe – there may be only one, elsewhere there will be two or more parties; in either case the fact remains that the party is not only the characteristic form of modern political organization, but also its hub. This is shown, in the communist one-party system, by the fact that the most important person in the state is the first secretary of the party; elsewhere the change is less clear and less complete but no less real. The essential point is that ultimate control, which during the period of liberal democracy was vested in parliament, has slipped, or

is slipping, from parliament to party – at different speeds
and by different routes in different lands, but everywhere
along a one-way road. Here again, a fundamental process
of modern mass society has been obscured by emphasis
instead on secondary phenomena, such as the threat to
parliamentary sovereignty from administrative law and
ministerial tribunals.

2

Political parties, it has been said, were born when the mass
of the population began to play an active part in political
life.[1] At first sight this statement looks like a paradox, or
even a dangerous half-truth; but we must not allow ourselves to be confused by nomenclature. It is true that we
find the word 'party' used to describe the factions which
divided the city-states of ancient Greece, the clans and
clienteles grouped round the condottieri of Renaissance
Italy, the clubs where deputies forgathered during the
French Revolution, the committees of local bigwigs in the
constituencies which ran elections under the constitutional
monarchies of the early nineteenth century, and the vast
countrywide party machines with their central offices and
salaried staffs which shape opinion and enlist votes in
modern democratic states. But if all these institutions have
one thing in common – namely, to capture power and
exercise it – in all other respects the differences between
them outweigh their similarities. In reality, political
parties as we know them are less than a century old.
Bagehot, writing in 1867, had not even a premonition of
the modern party system; what he envisaged was more like
a club than a modern party machine.[2]

It was, indeed, only in the last generation – in most

1. M. Duverger, *Les partis politiques* (4th ed., Paris, 1961), p. 466;
English trans., *Political Parties* (London, 1954), p. 426.
2. cf. W. Bagehot, *The English Constitution*, with an Introduction
by R. H. S. Crossman (London, 1963), pp. 39–40.

instances since the end of the Second World War – that political parties escaped from the limbo of extra-constitutional or conventional bodies, with no legally defined place in the system of government, and were explicitly admitted into the constitutional machinery. In England the change was registered by the Ministers of the Crown Act of 1937, which, by establishing the official position of the leader of the opposition, implicitly recognized and sanctioned the party system. In Germany the Fundamental Law of the Federal Republic – unlike the Weimar constitution, which still adopted an ambivalent attitude towards the party system – treated the parties as integral elements in the constitutional structure (Art. 21), and the Berlin constitution referred specifically to the tasks which fall to them under constitutional law (Art. 27). Similar provisions were incorporated in the constitutions of certain German *Länder* – for example, Baden (Art. 120) – in the post-war Italian constitution (Art. 49), and in the Brazilian constitution of 1946 (Art. 141).[1]

This legalization or constitutionalization of the party system was, of course, only a formal recognition of a situation that had long existed in fact. Nevertheless, it is only necessary to turn to the handbooks of constitutional law and political theory used in the inter-war period in England and elsewhere to see that it marked a real and substantial change.[2] In England, under the influence of Dicey, the interplay of parties was regarded as a useful

1. cf. G. Leibholz, *Der Strukturwandel der modernen Demokratie* (Karlsruhe, 1952), p. 16, for the Weimar constitution, ibid., p. 12.

2. An interesting example is H. J. Laski's *Grammar of Politics*, first published in 1925, since this book specifically set out to construct a new theory of the state, adapted to modern conditions. It is, however, only necessary to turn to the scanty passages (4th ed., 1941, pp. 264–6, 313–14, 318–24) in which the party system comes up for discussion to perceive that the really central issues are omitted. It is hardly an accident that the word 'party' is missing from the index; and a quick search of standard constitutional histories of the period would show that Laski was no exception.

'convention', helping the government to function smoothly, but not as an essential part of it, and as late as 1953 it was possible to write that 'the British party system is unknown to the constitution'.[1] In Germany, the existence of parties was ignored by Laband, and Jellinek specifically rejected the notion that they had any claim to a place in public law.[2] In France, where the modern party system was particularly slow to develop, the concept of the organized party was conspicuously absent from the standard textbooks of Barthélemy, Esmein, and Duguit, who admitted at most the existence of loose groups of deputies, brought together by similar tendencies and affinities, who might 'wish to maintain contact in order to concert their actions in a common direction in legislative and political questions'.[3] Today, these convenient fictions can no longer be maintained. We know, on the contrary, that the impact of organized parties has transformed not merely the infrastructure but also the substance of the parliamentary system, and that the part played by them is certainly no smaller than that of older organs of government, such as the monarchy or the cabinet. Today the British political scene is dominated by two great party oligarchies which have taken over and divided between them most of the sovereign powers Bagehot ascribed to the House of Commons. What we still think of as a parliamentary state has, in fact, become a party state, and the parties are now one

1. I. Bulmer-Thomas, *The Party System in Great Britain* (London, 1953), p. 3. R. T. McKenzie, *British Political Parties* (London, 1955), p. 4, also commented that 'despite their size and importance British parties are almost completely unacknowledged in law' and that there was 'no formal recognition of their role'.

2. cf. Leibholz, op. cit., p. 11.

3. J. Barthélemy, *Essai sur le travail parlementaire* (Paris, 1934), p. 91. In his standard textbook, *Le gouvernement de la France* (new ed., Paris, 1939), pp. 43–4, Barthélemy went out of his way to avoid using the word 'party', speaking only of 'groupes politiques . . . qui ne correspondent à aucune organisation dans le corps électoral'. Also cf. L. Duguit, *Traite de droit constitutionnel*, vol. II (Paris, 1928), p. 826.

of 'the most central and crucial of all the institutions of British government',[1] as, indeed, of government everywhere.

This change was the result of the appearance of a mass electorate which the old forms of political organization could not reach. It occurred, naturally enough, in different countries at a different pace, and its progress was affected in each different land by the pre-existing conditions. As already indicated, the United States, where conditions were more fluid and development less hampered by privilege and precedent, was ahead of Europe. In the United States manhood suffrage (for whites but not for Negroes) was already general by about 1825, and from approximately the same date mass immigration from Ireland, Germany, and Scandinavia built up a vast amorphous electorate. Except in the south, where before the civil war political power was lodged in the hands of a small stratum of wealthy planters, it was not long before the great families of the eastern seaboard, which had taken control in the revolutionary and post-revolutionary years, lost their pre-eminent position; and from the time of the election of Andrew Jackson in 1828 the outlines of the party machinery which was to dominate in the future – the machinery of bosses, managers and rings operating through graft, spoils and patronage to capture primaries, organize the 'ticket', and to manipulate committees and conventions – were already plainly visible.[2]

Characteristically, the change was resisted at first as 'contrary to republican institutions and dangerous to the liberties of the people', and was denounced as a 'Yankee trick' to prevent individuals from standing as candidates for Congress in their own right and deprive electors of

1. cf. D. Thomson in the *Survey of Contemporary Political Science* published by UNESCO (Paris, 1950), p. 546.

2. Its rise was described by M. Ostrogorski, *Democracy and the Organization of Political Parties*, vol. II (London, 1902), pp. 41 ff.

their freedom to vote for candidates of their own choice.[1] It also took some twenty-five years before the system was fully elaborated, and it only became continent-wide after the civil war had carried it into the south. But with the election of Harrison in 1840 and Polk in 1844 the American form of mass democracy had arrived, a third of a century or more ahead of the rest of the world. Polk was the prototype of the 'dark horse' candidate – the man on whom the masses could unite because he was sufficiently unknown or too colourless to arouse antagonism – but Harrison, epitomizing all the ideals of the 'log cabin' pioneers of the west, was swept into office, like Jackson before him, as what Max Weber was later to call the 'charismatic leader'. As for Martin van Buren, the organizer of Jackson's victory, he was the ancestor of a long dynasty of party managers and 'wire-pullers' with a world-wide progeny, among whom Alfred Hugenberg, the German press-lord who played so prominent a part in Hitler's rise to power, is perhaps the most notorious.

The transition from sedate liberalism, with its respect for birth, property, and influence, to mass democracy, which was an accomplished fact in the United States by 1850, was a far more hesitant process on the European side of the Atlantic. Here only the impact of industrialization in the period after 1870 was strong enough to override conservative resistance and carry the change through. The new political attitudes and methods manifested themselves first of all in England, immediately after the passing of the Second Reform Bill in 1867, though it was only after the passing of the Ballot Act of 1872, the Corrupt Practices Act of 1883, and the Third Reform Bill of 1884, which raised the electorate to around five millions, that democratization of the franchise could be said to have been secured. Perhaps the first clear victory for the new industrial democracy was the election of 1906, which, as Balfour

1. ibid., pp. 54, 66.

immediately perceived, inaugurated a new era.[1] In Germany, the decisive turning-point was the abrogation of the anti-Socialist laws in 1890.[2] Its immediate result was the rapid expansion of the Social Democratic party, founded in 1875, which now quickly drew ahead of all other parties, polling nearly one and a half million votes in 1890, over two millions in 1898, three millions out of an electorate of nine millions in 1903, and four and a quarter millions in 1912.

In Germany, as elsewhere in Europe, it was the socialist left wing that led the way in the development of new forms of political organization; with over a million inscribed members and a budget of over two million marks a year, the German Social Democratic party in 1914 constituted something not far short of a state within the state.[3] The bourgeois parties could only follow lamely after. Friedrich Naumann, appealing in 1906 for a revival of liberalism, was well aware that only a permanent, well-organized professional machine could bring it about; but his clarity of vision and purpose was exceptional, and the German middle classes were too divided socially to build the mass party for which he called.[4] The same was true in France. Here, indeed, the whole social structure, with its basis in a strong landowning peasantry and an extensive petty bourgeoisie, and its emphasis on regional differentiation and the antithesis between Paris and the provinces, was unsympathetic towards the rise of strong national parties. As late as 1929 the term 'party' was described as 'an agreeable fiction', so far as France was concerned; and even so distinguished an inter-war parliamentarian as André

1. Smellie, op. cit., p. 226.

2. cf. T. Nipperdey, 'Die Organisation der bürgerlichen Parteien in Deutschland vor 1918', *Historische Zeitschrift*, vol. CLXXXV (1958), p. 578.

3. cf. Duverger, op. cit., p. 90 (Engl. trans., p. 66).

4. cf. T. Schieder, *Staat und Gesellschaft im Wandel unserer Zeit* (Munich, 1958). p. 127; Engl. trans., *The State and Society in our Times* (Edinburgh, 1962), pp. 98–9.

Tardieu repudiated the notion of party attachment: 'I belong to none of those mystifications which people call parties, or leagues,' he said.[1]

Nevertheless France also was carried along by a development that was universal. As Maurice Deslandres wrote in a widely noticed article in 1910, 'worked upon by the new democratic ferment', the mass of the nation was rising up and establishing associations, leagues, unions, federations, committees, groups of militants, whose purpose was to activate political institutions and bring them, so far as possible, under their own tutelage. 'In the great unorganized homogeneous masses,' he said, 'a process of differentiation' was taking place, and in this way the country was 'becoming conscious of itself'.[2]

The event which, more than anything else, acted as catalyst in this process was the Dreyfus case. Overshadowing French politics between 1896 and 1899, the Dreyfus *affaire* discredited the opportunist bourgeois patriciate, which had monopolized power since the beginning of the Third Republic, and gave the left and the petty bourgeois left-centre an opportunity to play an active political role.[3] Hence in France it was in the first decade of the twentieth century that the new parties were organized: the Radicals in 1901, the *alliance républicaine et démocratique* the following year (though, characteristically, it was not until 1911 that the word 'alliance' was replaced by the word 'party'), the *fédération républicaine* in 1903, and the Socialist party (S.F.I.O.), formed by the amalgamation of a number of small existing rival groups, in 1905.

Although even now the formation of efficient party organizations, capable of disciplining the electors and

1. cf. R. von Albertini, 'Parteiorganisation und Parteibegriff in Frankreich, 1789–1940', *Historische Zeitschrift*, vol. CXCIII (1961), p. 594.

2. Deslandres's article, in *Revue politique et parlementaire*, vol. LXV, is cited by Albertini, p. 565.

3. cf. P. Miquel, *L'affaire Dreyfus* (Paris, 1961), pp. 9, 123.

controlling the deputies, was far from complete, the change was considerable. Its nature was indicated by two significant and representative statements, the one from 1900, the other from 1910.[1] 'If the electors are looking for direction,' wrote a competent observer at the earlier date, 'they will not find it in national organizations of a permanent character, putting before the country clearly defined courses of action; for such organizations do not exist. Hence each individual will vote without raising his eyes beyond the village pump. . . . And in parliament itself the situation is similar. There are no parties there; there cannot be. Each deputy has been elected separately; he arrives from his village with an essentially local programme. There is no flag for him to follow, no leader to rally and direct him.' By 1910 this was no longer the case. 'The word party, which formerly was used to designate an opinion,' it was pointed out, had now come to be used to denote 'an association founded to maintain that opinion'. It was true that, in France, 'the psychological factor of individualism was too strong for the parties to have the rigidity and precision of machines', but they were no longer simply organizations brought together intermittently on an *ad hoc* basis to fight particular elections.[2] As in Germany twenty years earlier, amateurish short-term electoral skirmishes were giving way to systematic long-term electoral campaigns; the old methods and the old machinery were no longer capable of coping with an electorate of many millions.[3]

3

What were the changes that were needed to meet the conditions of mass democracy, and how were they put

1. They are cited by Albertini, op. cit., pp. 566 and 567.
2. cf. L. Jacques, *Les partis politiques sous la IIIe République* (Paris, 1912), pp. 28 ff.
3. Nipperdey, op. cit., p. 579.

through? So far as England is concerned, the facts are reasonably well known and have been recounted in some detail, though most writers have tended to treat them as a process of continuous development and to slur over their revolutionary nature and revolutionary consequences. The starting-point was the Reform Bill of 1867 with its increase in the suffrage in the towns, and among the well-known milestones which followed were the organization of the radical 'caucus' in Birmingham by Schnadhorst and Chamberlain in 1873, its extension to other large cities, the formation of the National Liberal Federation in 1877, and Gladstone's Midlothian campaign of 1879. On the Conservative side these innovations were counterbalanced by the Conservative Working Men's Associations, the National Union of Conservative Associations, and the Primrose League, founded shortly after Disraeli's death in 1881.[1]

On the continent of Europe the process of renovation was carried through far less energetically than in England, but here also the necessity of underpinning the parties by widening their popular base could not fail to be recognized. Thus in Germany the Conservatives, who had hitherto largely dispensed with popular support because they were able to count on government backing, became from 1893 the organ of the Agrarian League, at the same time seeking a foothold among the artisans through the so-called *Bürgervereine*; while the Catholic Centre built itself up into a mass party through skilful manipulation of a variety of Catholic associations.[2] In France the Radicals tried to organize themselves on a nation-wide basis, by combining local committees into regional federations, with the party Congress at the head; but their success was

1. These developments were first analysed by Ostrogorski, op. cit., vol. I, pp. 161–272, and though the story has subsequently often been retold, his account is in many ways still unsurpassed. There is an appreciation of Ostrogorski and his work in M. M. Laserson, *The American Impact on Russia* (ed. 1962), pp. 473–84.

2. For further details, cf. Nipperdey, op. cit., pp. 581–90.

limited,[1] and in France it was only with the formation of the S.F.I.O. that anything like a mass party came into existence. Even so, it was a mass party without the masses.[2] So far as its organization was concerned, the S.F.I.O. conformed to the new model, but its actual membership in 1914 was only ninety thousand at a time when the German Social Democrats numbered a million. The first real mass party in France, with a membership reaching a million, was the Communist party; its phenomenal success, it has rightly been said, was almost certainly due more to its 'admirable system' of organization than to the attractions of Marxist doctrine.[3]

Four main factors distinguished the new forms of political organization. The first was a wide popular basis, or a mass membership; the second was permanence, or continuity; the third was enforcement of party discipline; and the fourth (and most difficult to attain) was organization from the bottom upwards instead of from the top downwards – in other words, control of policy by party members and their delegates instead of by a small influential clique in or about the government – the Carlton Club in London is the best-known example – or at the head of the party machine. All four points marked a radical break with the past. Earlier organizations had been largely intermittent; they had existed – like the Anti-Corn Law League in England, for example – to propagate a particular objective and had lapsed when it was achieved, or they had been called together to fight a particular election and had disbanded the day after the poll. In normal circumstances the smallness of the electorate meant that they were controlled by a few local bigwigs, usually the heads of county families, marked out by birth or wealth, who set themselves up with no further authorization as an *ad hoc* committee.

1. Albertini, op. cit., pp. 572–5; cf. also for a more general conspectus D. Thomson, *Democracy in France* (London, 1946), pp. 105–7.
2. Albertini, op. cit., pp. 592–3.
3. Duverger, op. cit., p. 22 (Engl. trans., p. 5).

None of these extra-constitutional organizations, as Ostrogorski has observed, entertained designs of making itself a permanent body, 'a regular power in the state'; none, in particular, set out to control the members of parliament or deputies whom it elected. The doctrine enunciated by Burke and Blackstone, according to which the deputy was the representative of the nation, not the mandatory of a party, and was consequently responsible only to his own conscience, was unquestioned in France and Germany as well as in England. With the extension and redistribution of the franchise all this was changed, and the main instrument of change was the device known as the 'caucus' – a word, significantly, of American origin, the adoption of which reflected the assimilation of American political ideas and practices. The caucus was the main political innovation of a new period; it provided, in Lord Randolph Churchill's words, 'undeniably the only form of political organization which can bring together, guide and direct great masses of electors'.[1]

As envisaged by its organizers, Schnadhorst and Chamberlain, and put into practice in Birmingham, the caucus was a party machine of a permanent character, built up out of cells in each ward or vestry, delegates from which formed the executive and general committee for the whole city, while the organizations of the different cities were linked together by the National Liberal Federation. Thus a machine was forged which, since it existed and functioned continuously and not merely at election times, was able to exert pressure on and even control members of parliament, and which because of its power could influence and sometimes even dictate policy. When in 1886 the caucus drove the independent radical, Joseph Cowen, out of public life, his comment was that what it wanted was a machine, not a man.[2] There is no doubt that in substance he was right. Hartington also complained that Chamber-

1. Ostrogorski, op. cit., vol. i, p. 275.
2. The Cowen case is discussed at length, ibid., pp. 231–42.

lain had organized an outside power to the belittlement
of parliament, and Harcourt told Morley that all that was
now expected of ministers was to swear loyalty to a creed
formulated by the Federation.[1] Nor was the wind from the
Tory quarter any less sharp than the gale from the radical
heights. What Chamberlain did to the Liberals, Lord
Randolph Churchill did to the Conservatives, installing
in place of the old lax methods and aristocratic cliques
'a new kind of plebiscitary caesarism, exercised not by an
individual but by a huge syndicate'.[2]

In principle, the emergence of the caucus marked a
radical breach with the past. In practice, it was otherwise.
The slowness and reluctance with which the bourgeois
parties adapted themselves to the condition of mass democracy is remarkable. Having advanced so far, they tended,
if anything, to draw back. The basic reason, without doubt,
was the unwillingness of the middle classes, with their
individualist traditions, to subject themselves to strict
party discipline, and the lack of a clearly defined class
interest to weld them together.[3] In addition, the party
leadership was adept at fighting back. In England both
the Liberal and the Conservative Associations were
brought with surprising speed under the thumb of the
party leaders,[4] and in France, Germany, and Italy the
outcome was much the same.

In France, the Radicals failed utterly in their efforts to
impose party discipline;[5] in Germany, as in England, the

1. Smellie, op. cit., p. 198.
2. Ostrogorski, op. cit., vol. 1, p. 282.
3. cf. Nipperdey, op. cit., p. 594, and Duverger, op. cit., p. 39
(Engl. ed., p. 21).
4. cf. Ostrogorski, vol. 1, pp. 302-4, 322-3. Instead of becoming the
suns of the party system, says Ostrogorski, the representative organizations became to all intents and purposes the satellites of the
leaders. This evolution has subsequently been analysed more fully
by McKenzie; cf. op. cit., pp. 584 ff., where the conclusions of his
study are summarized.
5. Albertini, op. cit., p. 578.

parliamentary leaders dominated the party congresses, arranged their proceedings in advance, and turned them into docile instruments of a governing clique.[1] Hence, although the tendency for the development of mass parties was everywhere strong, it was not until the appearance on the scene of the Socialist parties that the last obstacles were overcome. In the end, it was only fear of revolution and the growth of Communism that convinced the middle classes of the inadequacy of their traditional loose forms of organization and of the need to create mass parties; and the result was the emergence in 1932 of the National Socialist party – originally a right-wing splinter group, but now the petty bourgeois party *par excellence* – with 800,000 members and over thirteen and a half million votes.[2] In the meantime, on the opposite wing the German Social Democratic party from 1891, the British Labour party from 1899, and the French Socialist party from 1905 had systematically adapted and applied the principles and methods which the caucus organization of the 1870s and 1880s foreshadowed.

By comparison with the bourgeois parties, the strength of the Socialist parties lay in their firm social infrastructure. The same factors which led to the rise of mass democracy – namely, large-scale industry and urbanization –had brought about profound changes in capitalist society, and the rise of the Socialist parties marked the adaptation of politics to this fact. In the first place, the emergence of the factory or mill with thousands of workers on its payroll altered the structure of capitalism itself; it led, as contemporaries were well aware, to the displacement of industrial capitalism – of which the characteristic form was the independent family business – by finance capitalism, of which the American multimillionaire, John Pierpont Morgan, may be instanced as a typical figure. In the second place, it meant that the workers as a class tended

1. Nipperdey, op. cit., p. 585.
2. cf. Duverger, op. cit., pp. 90–1 (Engl. trans., p. 67).

increasingly to be reduced to the position of anonymous 'hands', unknown to employers they never saw, and that the division between those who owned and those who operated the machinery of production, hitherto glossed over by the prevalence of small factories in which the master and his employees worked side by side, became a basic element in society. Unlike the bourgeois parties, which professed to be 'national' parties representing all classes, the Socialist parties had from the start no hesitation in accepting this basic division; they were class parties representing a homogeneous class interest. The advantage, from the point of view of party organization, was immense. Above all else, the appeal to working-class interests brought for the first time a mass membership, either through direct adhesions or (as in England) through the support of the trade unions.

The phenomenal growth of the German Social Democratic party has already been instanced.[1] In Great Britain, by enrolling the unions, the Labour party already had 860,000 members by 1902. But it was not only a question of gross numerical strength. More important was the existence of an active, disciplined membership, organized from the centre and paying regular subscriptions. Here the Socialists were far ahead of the middle-class parties, which had difficulty in organizing their supporters as active party members, were largely dependent on such local initiative as might develop in individual constituencies, and relied for their finances less on regular subscriptions than on subventions from wealthy donors.[2] The

1. Above, p. 135.
2. The German National Liberals, for example, were able to organize at most fifteen per cent of their adherents; Nipperdey, op. cit., p. 596. For the finances, cf. Albertini, op. cit., p. 576. In 1907 Radical deputies and senators in France paid fr. 200, Socialists fr. 3,000. The subscription of the local Radical Committees – originally they paid none – was fixed at fr. 30; but in 1929 only 527 out of 838 had paid up. When the introduction of membership cards was discussed in 1912 it was protested that 50 centimes was too high a fee,

difference was plainly visible in France, where the Radicals, as late as 1927, had no clear idea of the size of their membership, whereas the Socialists supervised the membership through the central organization and collected party subscriptions through a central treasury, which distributed quotas to the local branches, instead of vice versa.[1] It was seen also in the growth of internal party organization – that is to say, of a salaried headquarters staff – in which the Socialists were also the pace-makers.[2]

The consequence of this close and effective organization was greater control. Instead of a loose association of committees, organized on a local or regional basis, which lacked cohesion and had little power to control the parliamentary leaders at the centre, the Socialist parties were unitary organizations, constituted on the principle of 'democratic centralism' and built up out of 'sections' which remained subdivisions of the whole.[3] There is no doubt that this type of organization made for greater cohesion and a greater measure of discipline. Whereas in the bourgeois parties it was the rule for the party to be dominated by the parliamentary group, all the Socialist parties adopted measures to ensure that the deputies were subordinated to the party and, in particular, to prevent them from asserting control either in the party Congress

---

and though formally accepted in 1913, they were not introduced, in practice, until 1923. For some data on Liberal and Conservative money-raising in England, cf. Smellie, op. cit., p. 198 : 'we make out a list of peers and M.P.s who may be asked to subscribe . . .' There were 114 of them and they were asked for '£500 apiece'. Also cf. McKenzie, op. cit., pp. 594–7.

1. Albertini, op. cit., pp. 575, 589.

2. Data for Germany in Schieder, op. cit., pp. 158–9 (Engl. trans. pp. 124–5); in France the Radicals only appointed a Secretary-General in 1929 (Albertini, op. cit., p. 579).

3. For the contrast between the section (or branch) – 'une invention socialiste' – and the committee (or caucus) – 'une type archaïque de structure' – cf. Duverger, op. cit., pp. 21–2, 37–9, 41–3 (Engl. trans., pp. 4–5, 20–1, 23–5).

or on the executive.[1] In France every candidate had to sign an engagement to observe the decisions of the national party congress, and the British Labour party insisted from the beginning that candidates must 'abide by decisions of the group . . . or resign'.[2] Thus the principle of the mandate, which the caucuses had tried with only limited success to enforce in the bourgeois parties, came into its own: it was, if we compare it with the classical theory of representation set out by Blackstone and Burke, one of the clearest signs how radically the impact of mass democracy had changed the political system.

## 4

The revolution in political practice outlined above is still for the most part an incomplete revolution. The United States, with its federal constitution and presidential system, has gone its own way, and the American political parties 'have had to eschew discipline, suppress doctrine and fragment power'.[3] Elsewhere the institutions which, considered theoretically, may be regarded as typical of mass democracy, are nowhere found in an undiluted form. Theoretically, for example, the Socialist parties are controlled by a democratically elected party congress, so constituted as to prevent the domination of the parliamentary leaders; but it is notorious that, in practice, the development of rigid party oligarchies has reduced the

1. The position is discussed by Duverger, op. cit., pp. 211–32 (Engl. trans., pp. 182–202); for France, Italy, Belgium, and Austria, cf. ibid., p. 222 (pp. 192–3); for Australia and Great Britain, p. 226 (pp. 196–7).

2. Albertini, op. cit., p. 500; McKenzie, op. cit., p. 387. In 1911, however, the British Labour conference decided that candidates should no longer be required to sign the pledge (ibid., p. 474).

3. cf. C. Rossiter, *Parties and Politics in America* (ed. 1958), p. 61 – a brilliant analysis of the salient differences of the American party system, which cannot be discussed in detail here.

control of the rank and file to nominal proportions.[1] In this respect, as in many others, the structural differences between the working-class and middle-class parties are in practice far smaller than at first glance might appear to be the case, and this is particularly evident where, as the German Social Democrats did in the Godesberg programme of 1959,[2] the former for tactical reasons have repudiated their class basis and set out, like their bourgeois counterparts, to establish themselves as 'national' parties. In practice, it is extraordinarily difficult, if not impossible, to determine exactly where the control of policy in any party – even a Communist party – lies at any particular moment.

These facts, and others like them, have made it easy to maintain the comfortable doctrine of constitutional continuity, to argue that, in spite of appearances to the contrary, the changes which have occurred during the past century have not affected the fundamental structure of government. Nevertheless, whatever point in the process we may currently have reached, it is clear that we are in the midst of developments leading away from the supremacy of parliament and towards some form of plebiscitary democracy, expressed in and through the party system.[3] Parliament today, it has been said, is little more than 'a meeting place in which rigorously controlled party delegates assemble together to register decisions already taken elsewhere, in committees or party conferences'.[4]

What has happened is that the place of parliament in

1. These aspects, as is well known, were examined at length by Robert Michels, *Political Parties. A Sociological Study of the Oligarchical Tendencies of Modern Democracy* (ed. 1962, first published in German in 1911), and do not require further discussion here.

2. cf. A. Grosser, *The Federal Republic of Germany* (New York and London, 1964), pp. 58–60, and A. J. Heidenheimer, *The Governments of Germany* (London, 1965), p. 66.

3. cf. E. Fraenkel, *Die repräsentative und plebiszitäre Komponente im demokratischen Verfassungsstaat* (Tübingen, 1958), p. 58 (for England, pp. 16–18).

4. Leibholz, op. cit., p. 17.

the constitution has shifted substantially, both in relation to the head of the government and in relation to the electorate. The change was initiated by Gladstone in 1879 when, in his famous Midlothian campaign, he appealed over the head of parliament to the electorate, thus 'removing the political centre of gravity from parliament to the platform'.[1] It was registered by Salisbury when he wrote in 1895, that 'power has passed from the hands of statesmen', and had already been foreseen by Goschen when he observed of the Reform Bill of 1867 that, through it, 'the whole centre of gravity of the constitution had been displaced'.[2] Since that time the process has gone forward, aided by the growing complexity of government and the highly technical nature of the decisions that have to be made. The result has been to place greatly increased power in the hands of the prime minister and his professional advisers. It is well known, for example, that the decision to proceed with the A-bomb was taken by Mr Attlee on his own intiative, without prior discussion in the cabinet, and was not revealed to parliament until the first bomb had been tested in 1952.[3]

Among the factors accelerating this process of concentration, one was the strain and stress of war which in England enhanced the personal power both of Lloyd George and of Churchill and led the latter, for the more vigorous prosecution of the war effort, to by-pass parliament and cabinet on a number of major questions of policy and administration. Another was the reform of the civil service by Lloyd George in 1919 and its centralization under the Secretary of the Treasury, who was made directly responsible to Downing Street. Its effect was to bring about 'an immense accretion of power to the prime minister', who now became 'the apex not only of a highly centralized

1. Smellie, op. cit., p. 193.
2. ibid., pp. 182, 192.
3. cf. J. P. Mackintosh, *The British Cabinet* (London, 1962), pp. 431–2.

political machine, but also of an equally centralized and vastly more powerful administrative machine'.[1] In the German Federal Republic under Adenauer this process was carried to the point at which the State Secretaries in the chancellor's office became the pivot of government, and during Adenauer's absences the actual direction of affairs devolved not on the vice-chancellor, but on the head of Adenauer's chancellery, the notorious ex-Nazi, Hans Globke. Not only was the collective responsibility of ministers undermined, but they were deprived of the control over their own departments provided for in the Basic Law.[2]

It is clearly impossible to predict how far the process of concentration will go, or what form of government may eventually emerge as a result of these and similar changes. But that does not diminish their impact nor make it less important to register their effects. If we try to summarize the changes as they appear today, without reference to their historical background, the following are probably the points which will stand out.[3] First, the position of the deputy, the representative or member of parliament, has altered in fundamental ways. Although lip-service is still paid to the theory which makes him the representative of the whole nation, bound only by his conscience, it is obvious that the actual position is very different. In reality, as M. Duverger has said,[4] 'members of parliament are subject to a discipline which transforms them into voting-machines operated by the party managers'. They cannot vote against their party; they cannot even abstain; they have no right to independent judgement on questions of substance, and they know that if they

1. For a summary of these developments, cf. Crossman, op. cit., pp. 48–51, 54–5.

2. cf. Grosser, op. cit., p. 35; Heidenheimer, op. cit., pp. 97, 101–2; P. H. Merkl, *Germany, Yesterday and Tomorrow* (New York, 1965), pp. 254–5.

3. For the following, cf. Leibholz, op. cit., pp. 16–27.

4. op. cit., p. 463 (Engl. trans., p. 423).

fail to follow the party line they can have no expectation of re-election. The one indispensable quality demanded of them, in short, is party loyalty, and the theory of classical representative democracy, that the electors should choose a candidate for his ability and personality, has ceased to count. From the point of view of the elector, the result by nineteenth-century standards is tantamount in many instances to disfranchisement; he can only vote for the nominees of the party or parties, none of which may represent his views, or not at all, and the complaints raised against the system when it first emerged in the United States are from this point of view fully justified.[1]

It is, however, from the point of view of parliament itself, and of the parliamentary system, that the consequences are most striking. The result of the changes of the last fifty years has been a steady and in some instances calamitous decline in its standing. With the disappearance of the solid core of independent and independent-minded members, parliament's role as a check and control on the executive has in the ordinary course of events become a fiction. It is also no longer, as in Bagehot's day, the place where ministries are made and unmade. 'Whether the government will go out or remain,' wrote Bagehot,[2] 'is determined by the debate, and by the division in parliament.' Today, since their results – even in a crisis of confidence like that in England in the summer of 1963 – are a foregone conclusion, parliamentary debates have lost their earlier constitutive character, and it is not surprising that they rarely arouse popular interest. If, formerly, major political issues hung in the balance and the fate of the government might be resolved by their outcome, at the present time, when the issues have been decided in advance in the inner party conclaves, speeches in parliament are no longer intended to sway the judgement of members, but are aimed at the elector outside parliament, with the object of impressing him and confirming his faith in the

1. cf. above, p. 133.        2. op. cit. (ed. Crossman), p. 73.

party. They are, in short, part of the barrage of propaganda directed at the electorate by newspapers, loudspeakers, television, and all other available methods of mass persuasion, but of all these various media they are the most antiquated and least effective.

The result has been to shift the emphasis away from parliament to the parties, on the one hand, and to the government, on the other. Armed with a mandate from the electorate, the government has little need to pay attention to parliament; the traditional view that the cabinet system enables parliament to control the government is very nearly the opposite of the truth.[1] Thus parliamentary elections tend to approximate more and more to plebiscitary acts; the electors, in other words, vote not for or against a particular candidate, but for or against a party programme and the leaders chosen by the party to execute the mandate. Where, as in Germany, the elector votes not for an individual but for a party list, this is even more obviously the case; the election of 1957 in the Federal Republic, for example, was in effect neither more nor less than a plebiscite for or against Dr. Adenauer. Thus elections tend to become popularity polls, and only the very naïve will be surprised if, as a result, the party machines – undeterred by the most unpromising material – seek to build up their chosen leaders into 'television personalities' and the like. Parties exist to secure power: it would be foolish to expect them to be squeamish about the means.

### 5

These facts, and the tendencies they reflect, have often been used to draw up an indictment of party government. That, it need hardly be said, is no part of my intention. All I have sought to do is to indicate by examples the

1. As was pointed out e.g. by W. I. Jennings, *The Law and the Constitution* (London, 1933), p. 143.

nature of the changes which have occurred as a result of the impact of mass democracy. The very fact that they are widespread changes, not confined to any particular country, indicates that they are part of a general historical process; and it is significant that the new type of party organization immediately took root in the emergent states of Africa.[1]

It is also evident that they are irrevocable changes, which reflect a basic alteration in the underlying social structure, and like all such changes they carry with them inherent dangers, or at least inherent problems. One is the possibility of government falling into the hands of a technically proficient but fundamentally cynical and self-centred party *élite*, a powerful apparatus controlled by a bureaucracy united by the same interests.[2] Another is the manipulation of the party machine by lobbies and pressure groups. In western Germany, in particular, there is concern at the extent to which the C.D.U. is exposed to the pressure of organizations representing business and other sectional interests.[3] But if party government, like all other political systems, is open to abuse, the remedy is not to decry the system but to improve its operation, above all by strengthening democratic control and counteracting the tendency, inherent in all political parties everywhere, to develop a rigid, top-heavy oligarchic apparatus. Those who rebel against the modern mass party and hanker for a return to earlier forms of representative democracy are indulging in a dangerous form of nostalgia; they ignore

1. cf. below, pp. 189–93.
2. H. Abosch, *The Menace of the Miracle* (London, 1962), pp. 226–7, points out that of the delegates to the Social Democratic Congress in Hamburg in 1950 only 8.2 per cent were working class, of the leaders of the party only 4 per cent. 'What immediately strikes one is the fact that a very high proportion of the delegates either owe their position to the Party or are directly employed by it. . . . Nominally, the Executive Committee is elected by the Party Congress; in practice the Congress is elected by the Executive Committee.'
3. ibid., p. 87.

the fact that the only practical alternative to the two-party or multiparty state, under present conditions, is the single-party state.[1]

The changes which in the last sixty years have brought the parties from the periphery to the centre of political life are not accidents which can be undone; they are part of the revolution which has given contemporary history a distinct character of its own and altered all its basic postulates. As Ostrogorsky was so quick to perceive, the advent of mass democracy disrupted the existing framework of political society. Today we are living in a new age of politics. If, throughout the contemporary world – in the western democracies, under the communist system, and now in the ex-colonial territories of Asia and Africa as well – highly organized parties are everywhere found occupying a central place in the political structure, it is because, under the conditions of mass society which have arisen since the end of the nineteenth century, the party is the only available means of articulating vast masses of people for political purposes.

1. cf. Leibholz, op. cit., p. 32.

# VI

## THE REVOLT AGAINST THE WEST

*The Reaction of Asia and Africa to European Hegemony*

'The problem of the twentieth century,' said the famous American Negro leader, William E. Burghardt Du Bois, in 1900, 'is the problem of the colour line – the relation of the darker to the lighter races of men in Asia and Africa, in America and the islands of the sea'.[1] It was a remarkable prophecy. The history of the present century has been marked at one and the same time by the impact of the west on Asia and Africa and by the revolt of Asia and Africa against the west. The impact was the result, above all else, of western science and industry, which, having transformed western society, began in an increasing tempo to have the same disruptive and creative effects on societies in other continents; the revolt was a reaction against the imperialism which reached its peak in the fourth quarter of the nineteenth century. When the twentieth century opened, European power in Asia and Africa stood at its zenith; no nation, it seemed, could withstand the superiority of European arms and commerce. Sixty years later only the vestiges of European domination remained. Between 1945 and 1960 no less than forty countries with a population of eight hundred millions – more than a quarter of the world's inhabitants – revolted against colonialism and won their independence. Never before in the whole of human history had so revolutionary a reversal occurred with such rapidity. The change in the position of the peoples of Asia and Africa and in their relations with Europe was the surest sign of the advent of a new era, and when the history of the first half of the twentieth century – which, for most historians, is still dominated by

1. cf. Colin Legum, *Pan-Africanism* (London, 1962), p. 25.

European wars and European problems, by Fascism and National Socialism, and by Mussolini, Hitler, and Stalin – comes to be written in a longer perspective, there is little doubt that no single theme will prove to be of greater importance than the revolt against the west.

1

It is, of course, true that the emancipation of Asia and Africa and the development of the European crisis went hand in hand. Among the factors which facilitated the rise of independence movements in Asia and Africa, we must include the weakening of the grip of the European powers, largely as a consequence of their own discords and rivalries and of the wastage of resources in which their wars resulted. From the time of the First World War the incipient nationalist movements in the non-European world profited substantially from the rivalries among the colonial powers, and the sudden collapse of the European empires after 1947 was to a large extent a consequences of external pressures and of the impact of world politics. In Asia neither the British nor the French nor the Dutch ever recovered from the blows inflicted by Japan between 1941 and 1945; while in Africa and the Middle East they were checked and forced into retreat by pressures from the United States – acting directly and through the United Nations – which had a strong anti-colonial tradition of its own and was unwilling to stand aside while colonialism drove the peoples of Asia and Africa over to the side of the Soviet Union.

Nationalism came to Asia a century later than it came to Europe and to black Africa fifty years later than to Asia. Two external events in the early years of the twentieth century were a powerful stimulus in its rise. The first was the victory of Japan over Russia in the war of 1904–5 – a victory hailed by dependent peoples everywhere as a blow to European ascendancy and proof that European arms were not invincible. Its impact was redoubled when, ten years later, the Japanese defeated the Germans in Shantung; and the successful campaigns of Kemal Ataturk against France in 1920 and Greece in 1922 were greeted in the same way as Asian victories over western military power. The second event was the Russian revolution of 1905 – a revolution which produced scarcely an echo in Europe but which, seen as a struggle for liberation from despotism, had an electrifying effect throughout Asia. The wave of unrest extended as far as Vietnam,[1] and its impact, in sparking off the Persian revolution of 1906, the Turkish revolution of 1908 and the Chinese revolution of 1911, and in the new impetus it gave to the Indian Congress movement in 1907, was such that its consequences in Asia have been compared with those of the French revolution of 1789 in Europe.[2] The result was that, by 1914, in most countries of Asia and the Arab world, but not yet in tropical Africa, there were radical or revolutionary groups ready to take advantage of the conflict between the European powers to secure concessions by threats or pressure or bargaining.

After war broke out the European powers themselves encouraged nationalist movements in colonial territories in order to embarrass their enemies. The Germans, for example, incited the nationalists of the Maghreb to take up arms against France, while the British and French with greater success stirred up Arab nationalism in Syria, in Mesopotamia and in the Arab peninsula against the Turks.[3] They were also forced by the pressure of events to make concessions to their own subject peoples. In India,

1. cf. D. G. E. Hall, *A History of South-East Asia* (London, 1955), p. 646.

2. cf. I. Spector, *The First Russian Revolution. Its Impact on Asia* (Englewood Cliffs, 1962), p. 29.

3. For German intrigue in North Africa, cf. F. Fischer, *Griff nach der Weltmacht* (Düsseldorf, 1961), pp. 146–7; G. Lenczowski, *The Middle East in World Affairs* (Ithaca, 1952), pp. 57–9, 73–7, recounts briefly British dealings with the Arab nationalists.

for example, the famous declaration by the British government on 20 August 1917, promising 'the gradual development of self-governing institutions', was a direct consequence of the Russian revolution which threatened to open the way for a Turkish and German advance on India at a time when the Bolsheviks were calling on the Asian peoples to overthrow the 'robbers and enslavers' of their countries. By the end of the First World War the cracks in the edifice of European imperialism in Asia and Africa were already assuming serious proportions, and there were limits, as the British found in Egypt after 1919, to what repression and military measures could achieve. Troops brought in from Syria broke the back of the Egyptian insurrection, but, as Allenby soon discovered, the problem of administering a restive country still remained. The troops could not be everywhere. Even when France, a generation later, diverted the bulk of its colonial army – twenty-five per cent of all French officers and forty per cent of the non-commissioned officers – to the struggle with the nationalists in Indo-China, it was as much as it could do to retain control of the big towns and main roads.[1]

The world war also helped in the dissemination of western ideas. War-aims propaganda could not be confined to Europe. Wilson's Fourteen Points. Lloyd George's declaration in 1918 that the principle of self-determination was as applicable to colonies as it was to occupied European territories, Lenin's denunciations of imperialism and the example of the Russian revolutionaries in declaring that the subject peoples of the Czarist empire were free to secede, all set up a ferment that was world-wide. Troops drafted to Europe from Indo-China by the French and from India by the British returned home with new notions of democracy, self-government, and national independence, and a firm resolve no longer to accept the old status of inferiority; among them was the future Chinese com-

1. cf. J. Romein, *The Asian Century. A History of Modern Nationalism in Asia* (London, 1962), p. 137.

munist leader, Chou En-lai.[1] A further factor fanning anti-European feeling was the failure of the European powers to carry out their wartime pledges. In the Near East and China the disclosure of the secret wartime agreements – the Sykes-Picot agreement between England and France to carve up the Ottoman empire and the agreement of February 1917 to hand over the former German possessions in China to Japan – discredited the western powers and provoked violent reactions. In China, the immediate outcome was the 'Fourth of May movement' of 1919, a decisive turning-point in the Chinese revolution.[2] In the Arab world the impetus to nationalism was equally strong. It was no accident that it was in 1919 that the Wafd party was founded in Egypt, or that in Tunisia it was in the same year that the Destour party, before coming out into the open as a legal organization in 1920, took shape as a clandestine underground group.[3] In Indonesia the same period saw the transformation of Sarekat Islam, founded in 1911 with limited and only semi-political objectives, into a mass movement demanding complete independence, to be obtained, if necessary, by force, and with a membership rising from three hundred and sixty thousand in 1916 to almost two and a half millions in 1919.[4]

The year 1919 also witnessed the convening of the first Pan-African Congress which met in Paris with the object of impressing on members of the Peace Conference the right of Africans to participate in government.[5] Its practical

1. cf. K. M. Pannikar, *Asia and Western Dominance* (London, 1953), p. 262.

2. cf. Chow Tse-Tsung, *The May Fourth Movement* (Cambridge, Mass., 1960), pp. 21 ff.

3. cf. N. A. Ziadeh, *Origins of Nationalism in Tunisia* (Beirut, 1962), p. 91.

4. cf. G. M. Kahin, *Nationalism and Revolution in Indonesia* (Ithaca, 1952), pp. 65–6.

5. cf. J. S. Coleman, *Nigeria: Background to Nationalism* (Berkeley, 1958), p. 188, and Legum, op. cit., pp. 28–9, 133–4. Legum designates the conference of 1919 as the second Pan-African Congress,

results, it is hardly necessary to say, were nil, for in tropical and central Africa, where most of the territories had only come under European domination after 1885, it was many years before the effects of European intervention in the form of roads and railways, industrial exploitation of mineral resources, the beginnings of western education and the like, produced substantial changes. In India, Malaya, and the Netherlands East Indies the First World War inaugurated a period of rapid economic development; in Africa south of the Sahara similar developments hardly got under way before the Second World War.[1] Nevertheless the Pan-African Congress of 1919, followed by others in 1921, 1923, and 1927, was indicative of the awakening which the ferment of the First World War stimulated and of the way ideas of self-government and self-determination were spreading. Every blow struck for independence reverberated over an ever-widening field, and there was a new sensitivity in each part of the dependent world to political developments in the others. The achievements of the Indian Congress were followed with lively attention, Gandhi's strategy of passive resistance was quickly adopted as a model, and similar organizations were built up in Africa and elsewhere as the hard core of revolt.[2] The Bolsheviks, who were aware of the revolutionary potentialities of Asia, did their best to keep up the ferment, and the Congress of the Peoples of the

---

since an earlier conference had been held in London in 1900; but this is contrary to usual practice, and most Africans have regarded it as the first in the series; thus, for example, Kwame Nkrumah, *Autobiography* (Edinburgh, 1959), p. 44.

1. cf. below, pp. 173–4, 190–91.

2. An example is the Sudan Graduates' Congress, founded in 1937. As Hodgkin points out, *Nationalism in Colonial Africa* (London, 1956), p. 146, 'the word "congress", here and elsewhere in colonial Africa, has obvious Indian associations'. Nkrumah, in his autobiography (p. vi) tells how, 'after months of studying Gandhi's policy', he came to the conclusion that 'it could be the solution to the colonial problem'.

East, which they organized in Baku in 1920, brought together delegates from thirty-seven nationalities.[1] In the Moslem world Pan-Islamic movements formed a link between countries as far apart as the Dutch East Indies, French North Africa and India, and facilitated cooperation between different nationalist groups.[2]

In this way the national movements of Asia and Africa gradually developed into a universal revolt against the west, a rejection of western domination which found expression in the Afro-Asian conference at Bandung in 1955. The Bandung conference symbolized the new-found solidarity of Asia and Africa against Europe; as Nehru said, it expressed the 'new dynamism' that had developed in the two continents during the preceding half-century.[3] Even as late as 1950 experienced western observers – Margery Perham, for example[4] – were expounding the comforting doctrine that, whatever the position might be in Asia, the day was still far distant when the African peoples would be capable of organizing independent states and, by implication, that imperial control and an enlightened paternalistic colonial administration would continue to be necessary for an indefinite period. No prediction could have been more fallacious. When the victory of Indian nationalism in 1947 and the collapse of European empires in Asia were followed by the failure of England and France in their war with Egypt in 1956, a new wave of nationalism pierced the barrier of the Sahara and swept across tropical Africa. After the Suez war of 1956 it was

1. For Lenin's policy towards Asia, cf. below, p. 215.
2. cf. H. A. R. Gibb, *Modern Trends in Islam* (Chicago, 1947), pp. 27–8, 32, 36, 119–20.
3. For the Bandung conference, cf. *Survey of International Affairs, 1955–1956* (London, 1960), pp. 59–65, where the main documentary sources are referred to.
4. cf. M. Perham, 'The British Problem in Africa', *Foreign Affairs*, vol. XXIX (1951), pp. 137–50. She thought it 'not a very bold speculation to believe' that the British colonial territories in Africa might 'become fully self-governing nation-states by the end of the century'.

clear – to governments in Europe, if not to intransigent minorities of white colonists in Africa – that the imperialist age had ended, and the European powers hastened, under pressure from outside and from within, to disburden themselves of colonies which had become a liability rather than an asset.

There is no doubt that external pressures, and the changing position of the European powers in the world, contributed to this great reversal. But pressures from outside, though they go far to explain the precipitate withdrawal at the end, only hastened a process of crumbling that had long been gathering pace; they could not have produced the result they did, if there had not been revolutionary nationalist movements within the colonial territories poised ready to take advantage of the difficulties in which the imperialist governments found themselves. More fundamental in the long run than the pressures resulting from the interplay of power politics were two other factors. The first was the assimilation by Asians and Africans of western ideas, techniques, and institutions, which could be turned against the occupying powers – a process in which they proved far more adept than most Europeans had anticipated. The second was the vitality and capacity for self-renewal of societies which Europeans had too easily dismissed as stagnant, decrepit, or moribund. It was these factors, together with the formation of an *élite* which knew how to exploit them, that resulted in the ending of European rule.

2

The history of the anti-western nationalist movements in Asia and Africa leads back, step by step, to the last two decades of the nineteenth century. In China it was the catastrophic defeat by Japan in 1894, and the threat of partition by the western powers which was its immediate consequence, that provoked a new nationalist response. In Egypt the anti-western movement was sparked off in the rising of Arabi Pasha in 1882, and began to make headway under the young khedive, Abbas II, who succeeded in 1892. In India the foundation of the Indian National Congress in 1885 paved the way for the quickening of the national consciousness after 1905. In the Ottoman empire the process of dismemberment at the Congress of Berlin in 1878 stirred into activity the patriotic movement of the Young Turks, who were to rise in revolution in 1908.

A later generation has seen in these reactions a turning of the tide. Earlier revolts – the Indian mutiny of 1857, for example, or the Senussi rising which followed the establishment of the French protectorate in Tunisia in 1881 – had been negative explosions of resentment and despair; they had represented the last convulsive resistance, hopeless if often heroic, of the old order. The new nationalist movements were in a different category. They looked to the future rather than to the past; and although in the first stage very disparate elements were found working together in their ranks, it is fair to say as a generalization that their object was not to throw off European domination by armed insurrection – a hopeless objective since, as the Boxer rebellion in China showed, fanaticism was no match for Maxim-guns – but to wear it down by erosion from within. Such a policy was, however, only practicable where social and other conditions were favourable, and it was no accident that the earliest nationalist movements occurred in countries which had a strong tradition of ancient civilization and an untarnished awareness of past achievement to fall back upon. They were also countries where western intervention had already shaken and weakened the old order. This was the case in India. It was also the case in Turkey, China, and Egypt, all of which had been compelled at an earlier date to throw open their doors to European commerce and which, as a consequence of the impact of European capitalism, had already passed through a generation or more of social ferment. Elsewhere conditions at the close of the nineteenth century were not yet ripe for the rise of nationalist movements. In tropical Africa, which was only caught in the European net in the last phase of imperial expansion after 1884, the impact of European capital investment was small until after the First World War and the forms of indirect administration in favour minimized social change. The exceptions were a few coastal areas, notably the Niger Delta and the Gold Coast, where European trading establishments were of long standing, and it is significant that it was here that the first stirrings of national consciousness were found. But in Africa south of the Sahara it was, in general, hardly before the 1930s that a conscious African political programme began to take shape, and most of the organized nationalist movements and parties of the region date from the Second World War or after.

The revolutionary movements which came to a head in the closing years of the nineteenth century, were a response to the disruptive effects of European intervention. When, between 1838 and 1841, Palmerston forced the Ottoman Sultan and the Egyptian Pasha to throw open their dominions to free trade, when by the treaty of Nanking in 1842 the same policy was forced upon the Son of Heaven, all three countries were thrust into an era of change with which none of the existing dynasties was equipped to cope. The stages in their disruption, the foreign loans, the deficits, the approaching bankruptcy, the dislocation of the economy through the influx of foreign goods, imperialist intervention to prop up the tottering régimes on which servicing and repayment of the debts depended, the crushing burden of taxation falling on a peasantry hard pressed already to earn a bare subsistence and on the edge of revolt, all these familiar aspects of a recurrent situation require no description. They created a ferment; they gave rise inevitably to resentment and hatred of the foreigner; they awakened doubts about the adequacy of traditional beliefs and institutions – of the accepted ethic of Islam or Confucius, for example, or of the traditional Chinese system of examinations – and awareness of the need for adaptation to the new world in order to survive; but they produced no coherent reaction. Hence these early movements have been called 'proto-nationalist' rather than nationalist.[1] They revealed the awakening of positive responses to the impact of the barbarians from the west, but mixed with more primitive reactions and not yet marshalled and organized into an effective movement which could seize and hold the initiative.

This was true of the revolt of Arabi Pasha in Egypt in 1881, the first reaction to the new situation. Four disparate elements crystallized round Arabi: small groups of liberal reformers, led by Sherif Pasha, who desired a western constitution and the regeneration they believed it would bring; Moslem conservatives, alarmed by the spread of Christianity and the religious laxity of the ruling class; disgruntled landowners, fighting to preserve their old fiscal privileges under guise of ridding the country of foreigners; and colonels smarting under the policy of military retrenchment enforced by the western powers.[2] In China, two decades later, the situation was much the same. Here the Manchu dynasty, already visibly in decline at the time of the Taiping rebellion half a century earlier, tried to exploit anti-foreign sentiment in order to regain support, while different groups of reformers sought ways out of China's dilemma. Those round Kang Yu-wei, loyal to the dynasty, fought to maintain the essential values of the Confucian system by bringing it up to date and reinterpreting the Confucian texts in the light of modern conditions, much as Mohammed Abdu in Egypt strove to revivify Islam by purging it of its reactionary elements;

1. cf. *New Cambridge Modern History*, vol. XI, p. 640.
2. cf. R. Robinson and J. Gallagher, *Africa and the Victorians* (London, 1961), p. 87; A. Hourani, *Arabic Thought in the Liberal Age* (London, 1962), p. 133. For a lively modern account of Arabia's revolt, cf. M. Rowlatt, *Founders of Modern Egypt* (Bombay, 1962).

others wished to take over western techniques, in the spirit of the great viceroy, Chang Chih-tung, without disturbing accepted beliefs and values; while the followers of Liang Chi-chao, convinced of the bankruptcy of Chinese tradition saw salvation only in a radical break with Confucianism.[1] Behind these and other intellectual groups stood a suffering peasantry and, as in Egypt, a class of young, ambitious officers, dissatisfied with the inefficiency, military and other, of the government.[2] It was a society in the throes of reconstituting itself, under external and internal pressures; but the dissident groups lacked unified leadership, coherence and clear objectives. The fruitlessness of attempting renovation within the existing system was exposed by the ill-fated Hundred Days reform of 1898; the disastrous consequences of turning popular discontent against the foreigner were shown by the outcome of the Boxer rising in 1900. Even the fall of the Manchu dynasty in 1911 seemed only to confirm the inability of China to adapt itself to the modern world, for amid the debris the conservative forces remained intact, and far from inaugurating a change for the better, the proclamation of the republic saw China falling apart among contending generals. The removal of the last of the Manchu emperors, Hsuan-tung, meant, in practice, only the destruction of the old Confucian conception of a unitary empire with one ruler at its head; failing any change in the social structure, it was unaccompanied as yet by constructive forces on a corresponding scale.

The fact remains that in China, as in Egypt, India, and Turkey, a revolutionary movement had been born, which, for all its confusion and the conflict of its disparate elements, was recognizably modern; and all these move-

1. There is a brilliant analysis of the intellectual currents in China at this period by J. R. Levenson, *Liang Ch'i Ch'ao and the Mind of Modern China* (2nd ed., London, 1959); cf. also the same writer's *Confucian China and its Modern Fate* (London, 1958–64).

2. cf. W. Franke, *Das Jahrhundert der chinesischen Revolution* (Munich, 1958), pp. 106–7.

ments reflected a common trend. If their immediate results were often negative, confirming the western belief that they were incapable of making the transition to modern conditions, all contained groups which looked to the future and which were determined to reconstitute their personalities on modern lines and regain their power by operating in the idiom of the westerners. And it was the irony of the situation that the European powers, once they involved themselves in Asia and Africa, could not help fostering and strengthening these elements. By propelling them into contact with a competitive economy and alien forms of government, they disrupted the existing balance upon which the stability of African and Asian societies rested; and their own active intervention, which soon followed, opened up an era of rapid social change which – no matter what line of policy they chose to follow – was bound eventually to turn against European rule. What is surprising is not the result but the speed with which – helped, as we have seen, by external events – it came about.


3

From the very beginning of the new imperialism in 1882 there were a few individuals with close knowledge of the orient who foresaw this result. Hart in China, the French consul in Cairo, warned the western governments of the dangers of the course they were embarking on and predicted the growth of an 'anti-European movement' 'destined to turn into fanaticism' and 'find expression in the wildest rage'.[1] At the time of the French advance in Indo-China in 1885 Jules Delafosse told the French Chamber that they were 'dreaming of a utopia' and that, before fifty years had passed, there would 'not be a single colony left in Asia'.[2] But it is not easy to see how or where the

1. cf. M. Bruce, *The Shaping of the Modern World* (London, 1958), p. 817; *New Cambridge Modern History*, vol. XI, p. 597.
2. Romein, op. cit., pp. 12–13.

European thrust, carried forward by its own inner logic, could have been voluntarily halted. Obsessed by their own rivalries, none of the European powers was prepared to stand aside while others extended their territories, or to withdraw and leave a void into which a potential enemy might move.

Against the gathering force of Asian and African nationalism the European powers found themselves, in the final analysis, with no effective defence. Considering their overwhelming superiority in arms and equipment, and their vast technological advantage, this was perhaps the most paradoxical aspect of the situation. The explanation, in the last resort, was demographic. How, for example, in the face of sustained civil disobedience, could Britain assure the long-term stability of its Asian possessions when, as we have seen,[1] the British in Asia numbered scarcely more than 300,000 out of a population of roughly 334 millions? Only where there was a substantial stratum of white settlers, as in South Africa and Algeria, was repression and the use of force an effective answer; the same factor and the advantage of a contiguous frontier was one reason – though it was not the only one – accounting for the relative success of Russian colonization in Asia.[2] But such conditions were the exception, and elsewhere the imperial powers were forced back on a policy of compromise and concession. Sometimes the concessions were the product of genuine enlightenment, for there were always elements in western society ready to raise their voice, on humanitarian and other grounds, against any form of colonial exploitation, and they were often able to bring effective pressure to bear; but in general they were the inescapable consequence of a situation which left the governing powers with no practicable alternative.

Though there were many local variants, the expedients

1. Above, p. 81.
2. For Russian policy in Asia, before and after the revolution of 1917, cf. below, p. 222; it cannot here be dealt with in detail.

to which the colonial powers resorted in order to preserve their supremacy followed a few simple patterns. First, there was the policy of indirect rule, support for princes and chiefs who were prepared in their own interest to collaborate with the occupying powers, which the British used in West Africa, the French in Indo-China and the Dutch in Indonesia. It had been an element of western policy ever since the European powers had thrown themselves behind the Manchu dynasty in its struggle with the Taiping rebels in China in the middle of the nineteenth century, and it implied for the most part maintenance of traditional societies as a bulwark against westernization and the disaffection it was liable to engender. Almost the reverse was the policy employed by the French in North Africa, where the danger seemed to come from conservative tribal and religious forces and where it therefore seemed sound tactics to build up a western educated *élite* of *évolués* which, it was hoped, would side with the progressive colonial power against reactionary nationalism. This was also, in effect, the assumption behind the Morley-Minto reforms of 1909 in India, which were postulated upon the existence of 'a class of persons, Indian in blood and colour, but English in taste, in opinion, in morals, and in intellect',[1] on which the government relied for support. Finally, there was the policy of offering internal self-government by instalments in the hope of staving off demands for independence – the policy of the Government of India Act of 1919 – or even of appearing to satisfy nationalist demands by granting quasi-independence but reserving essential rights – the solution sought by the British in Egypt and Iraq in 1922.

Over the short term these expedients often had a fair measure of success; in Iraq, for example, they ensured the maintenance of British influence until 1958. But it was also clear from an early date that they offered no solution and were only postponing the final reckoning. It has often

1. cf. *New Cambridge Modern History*, vol. XII, p. 215.

been said that the mistake of the imperialist powers was that the concessions they made to nationalist demands were 'always too small and too late to satisfy'.[1] This may be true so far as it goes; but if it is meant to imply that nationalism in Asia and Africa could have been satisfied by concessions short of full independence, it is necessary to add that this is an unverifiable assumption. There were certainly elements everywhere ready, not only for egoistic reasons, to cooperate with the imperialist powers, at least on a temporary basis; Dr Kwegyir Aggrey, the first African assistant vice-principal of Achimota College, for example – an outstanding personality, whom subsequent nationalist leaders such as Kwame Nkrumah looked up to with affectionate devotion – sincerely believed in cooperation.[2] But there is no reason to think that the situation could have been stabilized on this basis. The European powers, when they intervened in Asia and Africa, were caught in a dialectic of their own making; every action they took for the purpose of governing and developing the territories they had annexed made the maintenance of their own position more difficult, and there appears to have been no line of policy by which they could have escaped this fatal predicament. Nowhere, perhaps, is this more striking than in the history of British India after 1876. Here nothing is clearer than the ineffectiveness of what at the time seemed bold and radical changes of policy. Neither Lytton's conservatism nor Curzon's paternalism nor the liberalism of Ripon or Minto deflected Indian nationalism from its

1. ibid., p. 209.
2. 'He was extremely proud of his colour but was strongly opposed to racial segregation in any form. . . . Cooperation between the black and white peoples was the keynote of his message and the essence of his mission, and he used to expound this by saying: "You can play a tune of sorts on the white keys, and you can play a tune of sorts on the black keys, but for harmony you must use both the black and white" ' (Nkrumah, *Autobiography*, p. 12). Nevertheless for Nkrumah he was 'the most remarkable man that I had ever met and I had the deepest affection for him'.

course in any substantial way, and this was because ultimately nationalism was a response not to policies but to facts.

In these circumstances there is little point in discussing at length the different approaches of the different European powers to the problem of governing their colonial dependencies. At one stage the relative merits and demerits of 'association' and 'assimilation', of 'direct' and 'indirect' rule, and of other alternative systems, seemed to be a matter of immediate practical concern. Today it is evident that the distinctions for the most part were 'legal rather than practical'.[1] 'In practice association merely meant domination', and Léopold Senghor, the Senegalese leader, put his finger on the central defect of theories of assimilation when he said that what was needed – but not forthcoming – was 'assimiler, non être assimilés'.[2] If the immediate effect of indirect rule was to mitigate the impact of colonialism, it is also true that, by granting recognition to certain chiefs or princes only, and not to others, colonial governments tended over the longer run to create new, rigid patterns and to isolate the ruler, as an agent of the imperial authority, from his subjects.[3] Consequently the effect of 'colonial rule in any form or shape' was to cause 'a displacement of authority working against the traditional ruler'.[4] Where the western powers attempted to prop up existing dynasties as bulwarks against middle-class nationalism – for example, in Egypt – they only succeeded in discrediting them and involving them in the collapse of western positions; where they sought the cooperation of westernized *élites*, they weakened the only forces which had any lasting interest in maintaining European rule. Even on the lowest level of self-interest,

1. Hall, op. cit., p. 644.
2. cf. A. J. Hanna, *European Rule in Africa* (London, 1961), pp. 24–5.
3. cf. H. J. van Mook, *The Stakes of Democracy in South-East Asia* (London, 1950), p. 76.
4. F. Mansur, *Process of Independence* (London, 1962), p. 26.

the time was bound to come when westernized business-men in India or China or West Africa, who for a period might be ready to accept western rule for the commercial and industrial advantages it brought, would see more profit in displacing the foreigner and establishing a mono-polistic position of their own, and when westernized politicians would rebel against having to continue to share the spoils of office with the officials of the occupying power. But opposition to western imperialism was never, of course, simply an expression of crude self-interest. The desire for independence was pursued for the most part with unselfish devotion; and since European rule, however tempered by concessions it might be, necessarily implied dependence of some sort or other, the manoeuvres and contortions the imperialist powers went through until the very end, the offers and concessions and compromises they went on making in the hope of finding some formula which would save their own paramountcy while at the same time satisfying nationalist ambitions, were entirely unconvincing. At the same time they had to contend with the example of the 'white' dominions and *colons* who, however resolutely they might affirm their own superiority over the native population, were no less determined to assert their independent interests.[1] In the end, the differ-entiation between the 'white' and 'coloured' dependencies, so popular at the beginning of the twentieth century, became increasingly difficult to maintain; and once India, in 1947, had secured parity of treatment, the dam was irrevocably breached.

## 4

The same inner logic which carried European expansion to the bounds of the earth, not only invoked opposition and rebellion among the peoples brought under European rule, but also put new weapons in their hands. Both in

1. cf. above, pp. 68, 70 ff.

Asia and in Africa, European intervention had three necessary consequences. First, it acted as a solvent of the traditional social order; secondly, it brought about substantial economic changes; finally, it led to the rise of western educated *élites* which took the lead in transforming the existing resentment against the foreigner and foreign superiority into organized nationalist movements on a massive scale. All of these developments were necessary and unavoidable if the colonial powers wished – as naturally they did wish – to exploit their colonial acquisitions, or even, in most cases, if the colonies were to be made to pay their own way. Once the decision to intervene was taken, inaction was impossible; and action of any sort, even the loosest form of indirect rule, resulted in the crystallization of anti-western forces. What has been said of the Dutch in Indonesia applied to the colonial powers generally: 'the means chosen to defend the colonial régime . . . developed into one of the most potent of the forces undermining that régime.'[1]

The first of the consequences of European intervention – the disruption of the existing balance upon which the stability of the societies of Asia and Africa rested – was seen from an early date in India. Here, until experience of its results produced a reaction in the 1880s, English rule had deliberately undermined the old loyalties and sapped the power of the princes; it had operated as a great levelling force, breaking down the independent institutions of local political life, draining authority to a common centre, substituting British for Indian forms of law and administration, and weakening traditional religions, beliefs and habits.[2] Its impact on a simpler, less highly differentiated society is perhaps nowhere better expressed than in the measured, dignified statement made by the Brass chiefs

1. Kahin, op. cit., p. 44.
2. cf. E. Stokes, *The English Utilitarians and India* (Oxford, 1959), pp. 249 ff., 257 ff., 268 ff., 313 ff.

after the incident at Akassa in the Niger delta in 1895.[1] First, they stated, they had been prevented from making their living by selling slaves to Europeans as of old, which decision they had loyally accepted. Instead they turned to trade in palm oil and kernels. But then the British government opened the trade equally to 'white men and black men', to which also, 'seeing we could not do otherwise', they agreed. Finally, however, there came the Africa Company with a royal charter empowering it to do what it liked on the river Niger, and the result was that the tribesmen were driven away from the markets 'in which we and our forefathers had traded for generations', were forced to take out licences and pay heavy duties, so that – as they concluded – it was 'the same thing as if we were forbidden to trade at all'.

The report on the Brass disturbances provides us with an account, in the simplest of terms, of a process which took place wherever Europeans impinged on an alien people. What happened here, and at scores of similar points of contact in Africa, was the destruction of the economic substructure of tribal society, the erosion of the chiefs' authority, the transformation of the tribesmen, deprived of their traditional livelihood, into paid hands or servants of the foreigner, the loosening of social bonds, as they left their villages in search of alternative work, and ultimately their transformation into an urban and industrial proletariat. The obverse of this process, and usually its next stage, was the remoulding of the economy under the impetus of European enterprise. This was the second general consequence of European intervention. It proceeded at different speeds in different regions, but everywhere the two world wars were important turning-points. In colonial Africa, where, except in the mining areas of Rhodesia and Katanga, European investment was notoriously slow, it was only the Second World War that brought

1. cf. Sir John Kirk, *Report on the Disturbances at Brass* (Command Paper C.7977, Stationery Office, London, 1896), pp. 6–8.

to an end the stagnation of the preceding half-century. In Asia, on the other hand, the First World War gave a decisive impetus to the growth of modern industry. In China, the enforced inactivity of European merchants, whose home industries were concentrated on war production, created an opportunity for Chinese industry to go ahead in such fields as textiles, matches, and cement; cities such as Shanghai, Hankow, and Tientsin were industrialized, and new manufacturing centres grew up at important railway junctions such as Tsinan, Hsuchow, and Shihchiachuang.[1] In India, it was the deliberate policy of the British government to stimulate manufactures in order both to reduce the need for imports from the United Kingdom and to make India into a supply-base for Mesopotamia and other theatres of war.[2] The result was a vast impetus to the infant Indian iron and steel industries, which had only begun to produce between 1911 and 1914. At the same time, elsewhere in south-east Asia, the wolfram mines of Burma were developed until they produced one-third of the world's output, while the urgent needs of military transport necessitated a major expansion of rubber production in Malaya and the Netherlands East Indies. In Africa the results of the Second World War were similar. The disruption of older lines of supply, and the greatly increased demand for the strategic raw materials that Africa could provide, meant that the African colonies suddenly became of immense economic value.[3] The value of the Congo's exports increased fourteenfold, those of Northern Rhodesia ninefold, in a few years. In British West Africa the establishment of government purchasing agencies for vital products like vegetable oils and cocoa broke the hold the European trading companies

1. cf. Franke, op. cit., p. 145.
2. cf. *Cambridge History of the British Empire,* vol. v (Cambridge, 1932), p. 483.
3. cf. in summary R. Oliver and J. D. Fage, *A Short History of Africa* (London, 1962), p. 221.

had hitherto maintained over the peasant economies and paved the way for large-scale expansion, and the Colonial Development Act of 1940 – itself a direct result of war conditions – ensured that the impetus of the war years was not lost.

The consequence, first in Asia and then in Africa, was the growth of urbanization, of a factory working class that could be mobilized for political action, and of business communities which were wealthy enough to finance the independence movements. In the treaty-ports of China – notably in Canton and Shanghai – there grew up a wealthy Chinese commercial and industrial class, the so-called 'national capitalists' who placed themselves behind Sun Yat-sen, hoping for a stronger government which would defend their interests against their foreign competitors. Their type was C. J. Soong, father-in-law of Sun and of Chiang Kai-shek. In India, where the British economic impact was effective earlier, the typical figure was J. N. Tata, who opened the famous Empress cotton mill in Nagpur in 1887, and his sons who founded the Tata Iron and Steel Company in Behar in 1907. Here again European intervention had brought into existence a class which was vitally concerned to ensure its economic interests, and which threw itself behind Congress when it inaugurated the Swadeshi movement after 1905.

The rise of a new commercial and industrial middle class, with interests extending into finance and banking, was only one aspect of the process of rapid social regrouping touched off by the impact of the west. It was one of the most striking paradoxes of the situation that the colonial powers, having disrupted the existing social order, were compelled by their own requirements to create both a new class of leaders and the material and moral conditions which ensured the success of the anti-western revolt which they led. This was the third major consequence of western intervention. An educated Asian and African *élite*, adept in the techniques of western civilization, was a class that the imperial powers could not help creating, if only because they had an increasing need for cheap and plentiful clerical aid in the lower echelons of administration and business, and for skilled workmen in industry. The formation of the new nationalist *élites* was nevertheless a more complex process than is usually supposed, and it would be mistaken to think of it as simply the displacement of the traditional leaders by a newly emerging middle-class stratum. In Asia, at least, the new *élites* were not, for the most part, created at random out of a society diversified by the colonial impact – men who sprang up from hitherto politically inactive groups or classes – but were usually a section of the traditional ruling class, often a younger generation, which westernization had torn out of its traditional background.[1] In Africa, where Christianity acted as a democratizing influence, this was also true, but less regularly the case. Here there is more evidence of discontinuity in the traditional leadership, at any rate outside the Moslem areas. Thus leaders like Houphouet-Boigny and Sekou Touré are chiefs and sons of chiefs, but men like Nkrumah, Azikiwe, and Awolowo are usually accounted commoners, though it is noteworthy that Nkrumah, in his *Autobiography*, specifically refers to his high descent and his 'claim to two stools or chieftaincies'.[2]

Nevertheless the social upheaval created by the colonial impact was of decisive importance. Though the old ruling groups survived and provided many of the most notable nationalist leaders, westernization gave rise to a significant shift within their ranks, which brought to the fore those individuals who, often as a result of western education, were temperamentally able to adapt themselves to the new conditions. What was important, above all else, was their ability to cast aside their traditional class prejudices and work together with other groups, such as the lawyers and

1. This is illustrated with many examples by Mansur, op. cit., pp. 16, 21, 64, 162.
2. Nkrumah, op. cit., p. 21.

businessmen, who had formerly played no part in political life, but for whom westernization opened up new possibilities. The clearest example, but only one of many, is the collaboration between Liaquat Ali Khan, a wealthy landowner of royal lineage, and Jinnah, the son of a modest businessman.[1] It was this amalgamation, as a result of westernization, of elements from different social groups and classes that led to the formation of new *élites*, bound together, in spite of their disparate origins, by a determination to throw off foreign domination. Western education, besides its obvious effect of spreading the whole gamut of western ideas, from Christianity to Leninism, had two main consequences: first, it bred a growing class of discontented, educated or semi-educated Asians and Africans – the 'Westernized Oriental Gentlemen' (or, disparagingly, 'Wogs') of India and the 'Standard VII Boys' of Ghana and Nigeria – who were shut out from the better posts, reserved for Europeans, and often unable to find employment of any sort commensurate with their qualifications; and, secondly, it brought rapid and abrupt changes in the social balance, since, in a society in which barriers to social mobility were breaking down, the better-qualified elements with a western training, no matter what their origins, gradually displaced the old, less adaptable ruling class. Hence it is fair to say that the new *élite* assumed power because it was better able to represent the new pattern of social forces. This was a universal process, as visible in Indo-China under French rule as in India and Africa under British; and it occurred in much the same way in China, where the abolition of the traditional examination system in 1905 undermined the position of the gentry who for fifteen hundred years had been the pillars of the Chinese state.

The impact of European imperialism on Asian and African societies not only drove home the imperative need for change and pointed the way to modernization through

1. cf. Mansur, op. cit., p. 65.

the assimilation of European ideas, techniques, and institutions. It also made plain the necessity for new methods and strategies. Since it was more than doubtful whether the traditional societies of Asia and Africa, hierarchical and stratified, were capable of regeneration, the tendency, increasingly powerful as time passed, was to combine social transformation with political emancipation, for without social change the prospect of political emancipation was small. It was no accident that in China and the Ottoman empire, for example, almost the first step in the process of national revival was to get rid of the ruling dynasties, whose traditionalism and lack of adaptability were held responsible for the failure to hold the western barbarians at bay. The birth of nationalism can thus be viewed not merely as a reaction against western domination, but as the first step in the displacement of a traditional way of life, no longer in tune with modern conditions. Nehru, for example, has recounted that he worked for independence 'because the nationalist in me cannot tolerate alien domination', but that he worked for it 'even more because for me it is the inevitable step to social and economic change': in all his speeches on political independence and social freedom he 'made the former a step towards the attainment of the latter'.[1]

5

The development of the nationalist movements in Asia and Africa occurred in three stages. The first can be identified with the 'proto-nationalism' we have already considered.[2] It was still preoccupied with saving what could be saved of the old, and one of its main characteristics was the attempt to re-examine and reformulate the indigenous culture under the impact of western innovation.

1. cf. Jawaharlal Nehru, *An Autobiography* (London, 1936), p. 182, and *Toward Freedom* (New York, 1941), p. 401.
2. cf. above, p. 163.

The second stage was the rise of a new leadership of liberal tendencies, usually with middle-class participation – a change of leadership and objectives not inappropriately described by Marxist historiography as 'bourgeois nationalism'. Finally, there was the broadening of the basis of resistance to the foreign colonial power by the organization of a mass following among peasants and workers and the forging of links between the leaders and the people. Not surprisingly these developments proceeded at different paces in different countries, and could be complicated by the impact of an exceptional personality, such as Gandhi, who fitted uneasily into any recognized category of revolutionary leadership. They took place more slowly in countries such as India, which pioneered the revolutionary techniques, and more quickly in countries where nationalist movements developing after the process of decolonization had begun could benefit from the precedent and example of the older areas of discontent. In Burma, for example, nationalist developments which in India lasted for almost three-quarters of a century were telescoped into the decade between 1935 and 1945,[1] while in the Belgian Congo, less than four years before it became independent in 1960, Lumumba was still content to ask for 'rather more liberal measures' for the small Congolese *élite* within the framework of Belgian colonialism, and it was not until 1958 that he founded the first mass party on a territorial basis, the *Mouvement National Congolais*.[2] Nevertheless there is a clear pattern running through the nationalist movements, and the sequence observable in Asia and Africa seems in all essentials to be the same; in most cases, also, the three stages of development can be identified with the policies and actions of specific leaders.

The process of change is clearest in India. Here the

1. Mansur, op. cit., p. 83.
2. cf. Patrice Lumumba, *Congo, My Country* (London, 1962), p. 182; Lumumba's political evolution is discussed by Colin Legum in his foreword to this revealing book.

representative names are Gokhale, Tilak, and Gandhi, and the stages of development correspond fairly accurately to the three periods in the history of Congress: 1885–1905, 1905–19, 1920–47. In its earlier phase Congress was little more than a large-scale debating society of upper-class membership, content to pass resolutions proposing specific piecemeal reforms, and Gokhale, like other early Congress leaders, accepted British rule as 'the inscrutable dispensation of providence', merely asking for greater liberalism in practice and a larger share in government for educated Indians.[1] With Tilak, after his rise to prominence between 1905 and 1909, this upper middle-class reformism was abruptly challenged. Tilak rejected liberal reform under British overlordship, and demanded nothing less than independence; he also rejected constitutionalism and advocated violent methods. Yet on social questions Tilak was essentially conservative, while his nationalism – unlike that, for example, of the elder Nehru – was backward looking, postulated upon a purified Hindu ethic which he opposed to that of the west. Tilak, in fact, marked an intermediate stage – the stage of nationalist agitation on a relatively narrow middle-class basis, with the disaffected students as a spearhead and little effort at systematic mobilization of the masses.

What propelled the Congress movement into a new stage was the return of Gandhi to India in 1915, his assumption of leadership in the following year, the substitution for non-cooperation, which affected only a few special groups – lawyers, civil servants, teachers, and the like – of mass civil disobedience, which brought in the whole population, and the reorganization of Congress by the Nagpur constitution of 1920, as a result of which it became an integrated party with links from the village to

1. cf. P. Spear, *India, Pakistan and the West* (London, 1961), p. 200. Nehru, in his autobiography (e.g. pp. 48–9, 63–4, 137, 366, 416), has much to say on the middle-class bias of Congress at this time and later.

the district and province and thence to the top. This is not the place to discuss Gandhi's complex and in many ways enigmatic character. Over the long run it was perhaps his greatest achievement to reconcile and hold together the many disparate interests of which Congress was composed – a task it is highly improbable that anyone else could have accomplished. But there is no doubt that his outstanding contribution in the phase immediately following the First World War was to bring Congress to the masses and thus to make it into a mass movement. It was when Gandhi launched his first national civil disobedience campaign in 1920 that 'India entered the age of mass politics.'[1] He did not, of course, work single-handed, and the efforts of his lieutenants, particularly Vallabhai Patel and Jawaharlal Nehru, should not be underestimated. It was Patel, a superb political manager, who organized the Kheda and Bardoli campaigns which galvanized the peasant masses into action; it was Nehru who combated the right-wing elements in Congress and maintained the impetus to social reform without which popular support might have flagged.[2] But although it was the new radical *élite* which took in hand the task of organizing the masses politically, it is fair to say that it was Gandhi who made them aware of the importance of the masses.[3] One significant result was that a nationalist movement which had originated in Bengal and long retained a Bengali imprint spread throughout the whole sub-continent and became, except in areas dominated by the Moslem League, an all-Indian movement; another was that Congress, which at the time of the First World War was 'a floating but vocal *élite* with few real ties to its followers', had acquired by the time of the Second World War 'an effective organiza-

1. M. Weiner, *Party Politics in India* (Princeton, 1957), p. 7.
2. There are good assessments of Patel's and Nehru's roles in the movement in R. L. Park and I. Tinker, *Leadership and Political Institutions in India* (Princeton, 1959), pp. 41–65, 87–99.
3. cf. Mansur, op. cit., p. 71.

tional structure reaching from the Working Committee down through several levels of territorial organization to the villages'.[1]

The pattern we can trace in India can be seen, not without appreciable variations, in China. Here the three stages of nationalist development may be identified with Kang Yu-wei, Sun Yat-sen, and Mao Tse-tung, their sequence represented by the Hundred Days (1898), the revolution of 1911, and the reform and reorganization of the Kuomintang in 1924.

Unlike Kang Yu-wei, who hoped to reform China within the framework of the Manchu monarchy, Sun Yat-sen was a true revolutionary. It is true that, in 1892 or 1894, he had founded a reformist society, which aimed no higher than the establishment of constitutional monarchy; but after the disillusionment of 1898 and the bloody suppression of the Boxer revolt in 1900 Sun definitely threw over constitutional methods and in 1905 organized a revolutionary group which was the forerunner of the National Party, or Kuomintang. Its objects were essentially political – the expulsion of the Manchus and the establishment of a republic – and although as early as 1907 Sun made reference to the third of his famous three principles, 'the People's Livelihood' (*Min sheng chu-i*), social problems and particularly the agrarian question played little part in practice in his programme at this stage.

Sun was, in fact, a liberal and an intellectual, who believed that China's political salvation lay in the attainment of democracy on the western model; before 1919, he was not hostile to the western powers and was prepared to leave the unequal treaties intact. But the failure of the republic after 1911 showed the limitations of this 'moderate' approach. It also revealed Sun's essential greatness as a leader. In terms of actual achievement Sun counted for little during the first ten years of the republic;

1. cf. Park and Tinker, op. cit., p. 185.

he had difficulty in retaining a foothold in Canton and the principal role in the revolutionary movement appeared to be passing to the leaders of the Fourth of May movement. But Sun was one of those rare men – in this respect not unlike Gladstone – who became more radical with age. Disillusioned with the western powers, and stimulated by the nationalist enthusiasm of the Fourth of May movement and the workers' strikes which followed on 5 June,[1] Sun reorganized his party at the end of 1919, made contact with the Russian Bolsheviks, and set to work to revise his programme. From this time, Sun was a pronounced and open anti-imperialist, preaching passive resistance on the Indian model and a boycott of foreign goods. More important, he now placed the economic question at the head of his programme, allied himself with the Chinese communist party, which was busy under Mao Tse-tung organizing the peasants of Hunan, and carried through a major reorganization of the Kuomintang with the object of turning it into a mass party with a revolutionary army as its spearhead.

This reorganization of 1924 was a turning-point in the Chinese revolutionary movement. It marked the arrival of the third stage, namely the combination of nationalism and social reform and the broadening of the basis of resistance by the mobilization of the peasant masses. From this point, however, the revolutionary movement in China diverged from that of India. The death of Sun Yat-sen in 1925 meant that there was no one to hold together, as Gandhi did in India, the divergent elements in the national party; in China the businessmen, financiers and landlords on the right wing of the movement allied with the army under Chiang Kai-shek and turned against the communists and the left. The rest is well known. En-

1. For the Fifth of June, important as the first political strike by the urban workers in Chinese history, and as a link between the intellectual and working-class patriotic movements, cf. Chow Tse-tsung, op. cit., pp. 151–8.

couraged and financed by a group of Shanghai business-men, Chiang in 1927 liquidated all communists within reach, finally forcing the remnant to withdraw in 1934–5 to a remote area in the north-west where they were out of reach of the national armies. The Kuomintang itself, under the control of reactionary groups, put aside all thought of land reform, and gradually the initiative passed to the communists under Mao. Their strength lay in the fact that they did not shrink from social revolution. In his testament, composed a few days before his death, Sun Yat-sen had written that forty years' experience had taught him that China would only attain independence and equality when the masses were awakened.[1] Because Mao succeeded in translating this conviction into practice, it was he, rather than Chiang, who emerged as Sun's true heir. 'Whoever wins the support of the peasants,' Mao declared, 'will win China; whoever solves the land ques-tion will win the peasants.'[2]

In the agrarian revolution they launched in 1927 in the rural border areas of Kiangsi and Hunan and which ten years later they carried from their mountain retreat at Yenan into northern Hopei and Shansi, the communists provided the peasants with a leadership and organization without precedent in Chinese history. They organized local government by soviets, in which the poor and land-less peasants had the major voice; they distributed land taken from the landlords to this rural proletariat; they welded them into a revolutionary army waging guerrilla warfare against the privileged groups and classes. In short, they tapped the great human reservoir of China, and in this way they carried through an irreversible social trans-formation, which brought the work begun by Sun to its logical conclusion. 'The political significance of mass organization,' it has truly been said, 'was the primary factor

1. cf. Franke, op. cit., p. 208.
2. cf. Shao Chuan Leng and Norman D. Palmer, *Sun Yat-sen and Communism* (London, 1961), p. 157.

that determined the success of the communists and the failure of the Kuomintang.'[1]

It would take us too far to follow, even in bare outline, the course of development in other countries in Asia and in the Arab lands of the Middle East and North Africa. The picture they present would not differ greatly in substance, though in the case of the later nationalist movements, where the sequence tended to be telescoped and affected by external events, divergences might be considerable. In Indonesia, for example, the first two stages in the growth of nationalism went according to pattern, but the transition to the third stage – that is to say, the mobilization of the masses by a revolutionary social and economic programme – hardly got under way before internal developments were overtaken by the Japanese occupation of 1942–5. Hence it seems fair to say that it was the Japanese who propelled Indonesia into independence, or, at least, who accelerated what might otherwise have been a long and difficult process.

For this there were a number of specific reasons. In the first place, Dutch colonial practice hampered and retarded the growth of an Indonesian middle-class, and so – unlike India or China – there was no substantial capitalist or entrepreneurial element to underpin the revolutionary movement in its earlier 'bourgeois' phase.[2] This meant that the only possible basis for a successful Indonesian nationalist movement lay in the establishment of effective liaison between the intellectuals who made up the nationalist leadership and the Indonesian masses. Here again, however, conditions were unfavourable. Although the number of landless agricultural workers increased rapidly during the last decades of Dutch rule, a 'revolutionary agrarian proletariat', such as existed in China, did not materialize; the village community still provided basic

1. cf. Ping-chia Kuo, *China, New Age and New Outlook* (revised ed., Penguin Books, 1960), p. 63.

2. cf. Kahin, op. cit., pp. 29, 60, 471; Hall, op. cit., p. 661.

social security, even during the depression of the 1930s, and this fact continued to act as an effective brake on political restiveness.[1] Furthermore, the relatively late development of a conscious anti-Dutch nationalist movement – it was scarcely articulate before the members of the Indonesian students' union in Holland, founded in 1922, began to return to Indonesia towards the end of the decade[2] – meant that from the beginning it was involved in the ideological conflict unleashed by the Russian revolution of 1917. It was the infiltration of left-wing elements that impelled the early cultural and religious nationalist movement, Sarekat Islam, into politics and led it, in 1917, to demand independence.[3] But there was no body able, like the Indian Congress, to hold the disparate groups together, at least until independence had been obtained, and the dissensions among the nationalists proved disastrous and made it easy for the Dutch to intervene.

The consequence was that, after the suppression of the communist revolt in 1926, the nationalist movement was thrown on the defensive. The second stage came with the founding, in 1927, of the Persarikatan (later Partai) Nasional Indonesia, led by Sukarno – a national movement founded deliberately on the model of Gandhi's non-cooperation campaign, which sought to bring together all the existing nationalist groups in one organization. If the leader of Sarekat Islam, Tjokro Aminoto, may be compared with Gokhale in India, then it may be said that Sukarno corresponds to Nehru and Jinnah. But although the P.N.I., under Sukarno, gave the nationalist movement a unity it had never previously possessed, the lack of a solid basis, in the form of a spontaneous revolutionary movement among the peasants, made it difficult to withstand

1. cf. Kahin, op. cit., pp. 18–19.
2. For the students' organization, Perhimpoenan Indonesia, cf. ibid., p. 88. Hatta and Sjahrir returned from Holland in 1932. Sukarno, trained as an engineer at the Bandung Technical College, did not belong to this group.
3. cf. above, p. 157.

Dutch countermeasures. By imprisoning the leaders – Sukarno was deported from 1933 to 1942, and soon followed by Hatta, Sjahrir, and other dynamic nationalist leaders – and by breaking up the trade unions after 1929, Dutch policy scored a fair measure of success. Attempts at organizational contact with the mass of the peasantry were almost completely frustrated, and the nationalist leaders were never, during the course of Dutch rule, able to come into sufficient contact with the peasantry to organize it effectively into the nationalist movement, which thus remained dependent on the white-collar workers, students, teachers and the like.[1] Without the organized backing of the peasant masses, however, the nationalist movement stood little chance of success against the repressive power of the Dutch. Hence the Japanese invasion, which broke Dutch power, was a turning-point. But it is also true that the Dutch, by welding together the peoples of various tongues and cultures who inhabited the Indonesian archipelago, helped to turn what had originated as Javanese patriotism into an all-embracing Indonesian nationalist movement. A further factor was the high degree of religious homogeneity that prevailed in Indonesia. As the nationalist movement spread out from its original base in Java, the parochial tendencies and local patriotisms that might otherwise have been strong among the peoples of the other islands were counteracted by a sense of solidarity springing from common adherence to Islam.[2]

The nationalist movement in North Africa also owed its early impetus to Islam and its development was roughly simultaneous with that in Indonesia. In Tunisia, for example, the old Destour, or Constitutional party, founded in 1920 by the Islamic reformer, Shaikh Abdul-Aziz ath Tha'alibi, with a programme of administrative reform in cooperation with France, was superseded after 1934 by Bourguiba's Neo-Destour, a radical and secular

1. cf. Kahin, op. cit., p. 63.
2. ibid., pp. 37–8.

mass party in many ways parallel to Sukarno's P.N.I.[1] And just as the Japanese occupation made it possible for the Indonesian independence movement to come out into the open, so in North Africa the presence of Anglo-American troops after 1942 made possible the transformation of the more rudimentary pre-war Moroccan political movements, the *Comité d'Action Marocaine* (1934–7) and the *Parti National pour la Réalisation du Plan des Réformes* (1937–9), into the more widely based Istiqlal, or Independence party, in 1943.[2]

In tropical Africa, also, the Second World War was a decisive turning-point. In the French colonies, in particular, the 'free French' had to promise substantial changes in order to win the support of the native population against Vichy. In other respects, however, the development of nationalism in tropical Africa followed a somewhat divergent course. In Africa north of the Sahara, as in Indonesia, the demand for independence grew out of earlier conservative Islamic movements, and the first reactions to the west were touched off by intellectuals who wished, as in China and India, to defend a cultural heritage threatened by disruption from without. In central Africa an intelligentsia of this type was lacking. There is no African Gandhi, no African Sun Yat-sen.[3] The early intellectuals, Garvey, Du Bois, and Blyden, were West Indians concerned – as Nkrumah later complained[4] – with '*black* nationalism as opposed to *African* nationalism'. In central Africa, therefore, the cultural counter-revolution was a product, rather than a cause, of the development of a self-conscious nationalist movement. The reason was that

1. cf. C. A. Julien, *L'Afrique du Nord en marche* (Paris, 1952), pp. 79 ff.; F. Garas, *Bourguiba et la naissance d'une nation* (Paris, 1956), p. 78; for Shaikh Tha'alibi, cf. Ziadeh, op. cit., pp. 98–102.

2. 'La présence américaine exalta le nationalisme'; Julien, op. cit., p. 342; cf. also T. Hodgkin, *African Political Parties* (London, 1961), p. 52.

3. cf. Hodgkin, *Nationalism in Colonial Africa*, p. 179.

4. cf. Nkrumah, *Autobiography*, p. 44.

Africans had no single comprehensive civilization and no common background of written culture to look back to. In this respect tropical Africa was more like Indonesia than India or China. It contained a multiplicity of peoples at very different levels of social life, and the object of the emerging nationalist leaders could not be to return to a past which was tribal and ethnic, but aimed rather at the creation of a new African personality. On the whole, therefore, African nationalists were not 'cultural nativists',[1] and the reaction against western civilization, which accompanied the rejection of western political domination in Asia, was never very strong in Africa. As Nkrumah wrote in 1958, it was the west which 'set the pattern of our hopes, and by entering Africa in strength . . . forced the pattern upon us';[2] and it is within this pattern that African nationalism has evolved.

With these exceptions, however, it is fair to say that the African response to alien rule and to the stimulus of westernization has followed 'a historic pattern'.[3] Here again, it is not difficult to trace three distinct stages of development. In the Gold Coast they were represented by the Aborigines' Rights Protection Society, the United Gold Coast Convention, and the Convention People's Party and identified with the names of Casely-Hayford, Danquah, and Nkrumah.[4] In Nigeria the pattern is more complex, for here the situation was complicated by persistent regional and tribal divisions and the strength of Islam in the north; but here also there is a clear line of development leading from the Nigerian National Democratic Party, founded in 1923 under the leadership of Herbert Macaulay, to the National Council of Nigeria and the Cameroons (1944), in which Azikiwe was the leading

1. Coleman, op. cit., p. 411.
2. cf. *Foreign Affairs*, vol. XXXVII (1958), p. 53.
3. Coleman, op. cit., p. 409.
4. cf. D. E. Apter, *The Gold Coast in Transition* (Princeton, 1955), pp. 35–7, 146, 167 ff.; F. M. Bourret, *Ghana. The Road to Independence, 1919–1957* (London, 1960), pp. 40, 54–5, 61–2, 69, 166, 173 ff.

figure, and then to the Action Group founded by Chief Obafemi Awolowo in 1951. The N.C.N.C. and the Action Group are often regarded as parallel organizations, the one deriving its strength from the eastern, the other from the western region; but, in fact, there is little doubt that the Action Group represented a more advanced form of political organization with a collegial leadership, modern techniques of campaigning and a clearly formulated programme. It also took a more uncompromising stand on the question of independence. The N.C.N.C., on the other hand, was not a mass movement – until 1952 there were no individual members – and failed to secure the adhesion either of the Youth Movement or of organized labour. Moreover, its original programme, as formulated in 1944, looked no further than 'self-government within the British Empire', and attempts after 1948 to propel it in the direction of militancy produced a reaction which led to a period of inactivity. It is fair to say, therefore, that the founding of the Action Group in 1951 marked the opening of a new phase.[1]

What we see, both in the Gold Coast and in Nigeria, is a characteristic evolution, from loose and often informal associations for reform within the existing colonial system, through middle-class parties with limited popular contacts, to mass parties which mobilized support by combining national with social objectives for the attainment of which the whole people could be stirred into action. This evolution is clearly parallel to that which, for the most part, had already occurred in Asia; indeed, it has been said that, with the founding of the National Congress of West Africa in 1920, there began in Africa the period which India had entered towards the end of the nineteenth

1. In the view of Coleman, *Nigeria: Background to Nationalism*, p. 350, the Action Group 'differed from all previous Nigerian political organizations'. For its pressure for independence, cf. ibid., pp. 352, 398, and for the more restricted objectives of the N.C.N.C., ibid., pp. 264–7. For the setback to the Zikist movement and the consequent decline of the N.C.N.C. around 1950–1, cf. ibid., pp. 307–8.

century and left in the years immediately following the First World War, and that the foundation of the U.G.C.C. and the N.C.N.C., in 1947 and 1944 respectively, started British West Africa on the road travelled by south-east Asia in the two decades of the inter-war period.[1] There are also clear parallels between the evolution of African political parties and the movement towards mass democracy which had begun, as we have seen,[2] three or four decades earlier in Europe. But the movement proceeded further and more logically in Asia and Africa, because there the development of mass parties was not hampered by the survival of earlier traditions of parliamentary government. Nevertheless it could only be carried through by new leaders less inhibited both in their relations with the colonial government and in their social outlook than the older leadership. As Nkrumah put it, 'a middle-class *élite*, without the battering-ram of the illiterate masses' could 'never hope to smash the forces of colonialism'.[3] In other words, social revolution was the necessary counterpart of national emancipation; only in this way and through the strict discipline of tightly organized national parties could a mass resistance be built up, against which the colonial governments would ultimately be helpless.

Only a brief summary is possible of the steps by which this transformation took place. Their background was the period of rapid economic and social change during and after the Second World War to which allusion has already been made.[4] Of this the most spectacular aspect – parallel in many ways to what was going on simultaneously in Soviet Asia – was the growth of towns; and the new towns generated both a social life of their own, unlike any that had previously existed in Africa, and a spirit of African radicalism which provided a ready-made material for the new generation of nationalist leaders, of whom Nkrumah is perhaps the typical example. Elisabethville almost

1. cf. Mansur, op. cit., p. 56.      2. cf. above, p. 133 ff.
3. *Autobiography*, p. 177.           4. cf. above, p. 173.

trebled in population between 1940 and 1946; Bamako doubled and Leopoldville more than doubled in the same short span of time; Dakar rose from 132,000 in 1945 to 300,000 in 1955.[1] Four main consequences ensued. First, the towns threw up a new stratum of tough, emancipated, politically active men, ready to follow a bold leadership which knew where it was going. Secondly, they provided a mass audience. Thirdly, they acted as new focuses of national unity, which cut through tribal divisions and formed an urban network binding together Africa's scattered rural communities. And, finally, the tremendous improvement in communications which economic progress necessitated enabled the leaders to forge organizations which covered the whole country.

As in Indonesia, it was the return from abroad of a new generation of leaders, schooled in politics, confident of their ability to handle western political techniques, and aware of the potentialities of the new situation, that made it possible to exploit these changes. The older generation was hampered by a sense of insufficiency. As one of them confessed during the constitutional debate in the Gold Coast in 1949, under colonial government their limbs had become 'atrophied through disuse' – 'we want faith and confidence in ourselves'.[2] They were also chary of seeking popular support, conscious that the political mobilization of the masses would weaken their own position. As Nkrumah scornfully remarked, 'the party system was alien to them', and he recounts how, when he took up his duties as general secretary of the U.G.C.C. in 1948, only two branches had been established 'and these were inactive'.[3] The return of Nkrumah from England in 1948

1. For these and other figures, cf. Hodgkin, *Nationalism in Colonial Africa*, p. 67. There are figures for the Gold Coast, based on the 1931 and 1948 census, in Apter, op. cit., p. 163. In this period Kumasi more than and Accra and Sekondi-Takoradi almost doubled in population.
2. cf. Apter, op. cit., p. 178.
3. *Autobiography*, pp. 57, 61.

thus marked a turning-point in Gold Coast politics, just as the return of Azikiwe to Nigeria in 1937 had opened a new period.[1] Like Azikiwe, Nkrumah realized that 'there is no better means to arouse African peoples than that of the power of the pen and of the tongue'.[2] His *Accra Evening News* performed the same function of inflaming racial and national feeling in Ghana as Azikiwe's *West African Pilot* did in Nigeria. At the same time – again like Azikiwe – he threw himself with intense energy into touring the countryside, addressing meetings, issuing membership cards, collecting dues, founding branches. Nkrumah himself has told how, within six months of arriving in the Gold Coast, he established five hundred branches of the U.G.C.C., how this enlistment of the rank and file alienated the Working Committee of the U.G.C.C. – 'it went completely against their more conservative outlook' – and how, when the latter refused to endorse his policy of 'Positive Action', he broke away and formed the Convention People's Party.[3]

The C.P.P. was from the first a mass party, but it was not merely a mass party, for, as Nkrumah said, 'mass movements are well and good, but they cannot act with purpose unless they are led by a vanguard political party'.[4] Nevertheless its victory in 1956 was due to its organization of the masses and to the strict discipline imposed on its members; it 'marked the ascendancy of an egalitarian, nationalist, mass party over a traditionalist, regionalistic, and hierarchical coalition'.[5]

The success of the C.P.P. in Ghana is only one of the more striking examples of a policy which other leaders were applying elsewhere in Asia and Africa. Trained in the United States, London, Paris, and sometimes in Mos-

1. For 'Zik', cf. Coleman, op. cit., pp. 220–4.
2. cf. N. Azikiwe, *Renascent Africa* (Accra, 1937), p. 17.
3. *Autobiography*, pp. 61, 79, 82, 84.
4. ibid., p. vii.
5. Mansur, op. cit., p. 88.

cow, they built up mass parties on the model of what they had observed in the west, with a pyramid of units running from local branches to national conferences, with a central office and a permanent secretariat, with their own newspapers, emblems, flags and slogans, and with cars, helicopters, loud-speaker trucks and all the other paraphernalia of political organization and propaganda. This was the type not only of the Convention People's Party in Ghana, but also of the Action Group in Nigeria, of Julius Nyerere's Tanganyika African National Union, of the *Rassemblement Démocratique Africain* and the *Bloc Populaire Sénégalais*. Their leaders knew, as Nkrumah recorded in his autobiography, 'that whatever the programme for the solution of the colonial question might be, success would depend upon the organization adopted'.[1] They were right. It was this perception that distinguished them from the earlier generation of nationalist leaders and enabled them to mobilize the forces which the impact of westernization had released in Asian and African society. On the whole, we can fairly say that those who mobilized the new social forces succeeded, those who held back and fought shy of mass agitation and social action did not. In essence, it was because it failed to cope with the agrarian problem and thus to meet the basic needs of the people that the Kuomintang missed its opportunity in China and was superseded by the Chinese Communist party under Mao Tse-tung and Chou En-lai. In India the course of events was difficult because Congress, though originating like the Kuomintang in the middle classes, made contact with the peasantry and, through the organizing genius of V. J. Patel, built a party machine which mobilized the masses, in the countryside as well as in the towns, behind the struggle for independence until it was won. In the end, the revolt against the west, both in Asia and in Africa, merged into a greater revolt still – the revolt against the past. Political independence, as Nkrumah said,

1. Nkrumah, *Autobiography*, p. 37.

was only 'the first objective';[1] what gave it strength, and won it overwhelming popular support, was the determination to use independence to build a new society designed to serve the needs of the people in the modern world.

6

No one who has studied the successive stages in the development of the nationalist movements in Asia and Africa can seriously doubt the influence exerted by western political practice and example. But we must be careful what corollaries we draw from this fact, and particularly careful before we accept the conclusion, common among western political commentators, that the impact of Europe was the catalyst that brought about the resurgence of Asia and Africa. As Sir Hamilton Gibb has written, the outward effects of the world-wide extension of western technology and skills are so obvious that it is easy to assume a parallel extension of western thought; but such an assumption would be 'quite unjustified'. In reality, 'the thought-forces now operating in the Muslim world are forces which have been generated within the Muslim community', even though their emergence has been due very largely to the impact of the west and the trend of their development has been partly determined by western influences.[2]

What the west provided, in the first place, was a motive: that is to say, Asians and Africans reacted against European domination, against their relegation to the status of inferior races, against what they regarded as exploitation for the benefit of European interests. It also provided the means and created the conditions for successful revolt. It was soon obvious that the traditional societies of Asia and Africa, even a state which had been as powerful and expan-

1. Nkrumah, *Autobiography*, p. vii.
2. cf. Gibb, *Modern Trends in Islam*, p. 109.

sive as the Ching empire of China, were no match for the European conquerors with their massive armaments and their new technology. The impact of Europe drove home the imperative need for change, the brutal realization that the only alternative to modernization was to go under. At the same time it pointed the way to modernization through the assimilation of European ideas, techniques and institutions, and facilitated this process by weakening the foundations of traditional societies. Hence it is often said that it was by exploiting European ideas of self-determination, democracy, and nationalism and by adopting the advanced processes of western industrialism and technology that Asians and Africans raised themselves from subjection to independence: they took over weapons forged in Europe and turned them against the European conquerors.

There is, of course, much truth in this analysis. But it is also true that the current tendency to treat westernization as the key to the revival of Asia and Africa leaves out some relevant facts. The more we know of Asian and African societies before the advent of the Europeans, the clearer it has become that they were neither stagnant nor static, and it would be a mistake to assume that, but for European pressures, they would have remained anchored in the past. In the Arab world, for example, the Wahhabite movement in the eighteenth century was clear evidence of spontaneous renewal. Japanese society was in the throes of change long before the arrival of Perry in 1853, and in China also an explosive process of social adjustment was under way by the beginning of the nineteenth century.[1] In any case, contact with Europe, though it may have created the conditions and provided the means, did not account for the will to secure independence. The transformation of Asian and African society by western industry and

1. cf. H. A. R. Gibb, *Studies on the Civilization of Islam* (London, 1962), p. 327; R. F. Wall, *Japan's Century* (London, 1964), pp. 6 ff.; Ping-chia Kuo, op. cit., p. 23.

western technology was a major factor in the situation; but it would not by itself have restored them to an independent position in the world unless it had been accompanied by other forces which did not stem from the west. These forces also played their part in the political reawakening.

Among them perhaps the most important was the determination of Asians and Africans to maintain, or reshape, or, where necessary, create their own 'personality'. At some periods, particularly in countries where Hindu or Moslem tradition was strong, this took the form of a flight into the past. On the whole, however, this conservative and largely sterile reaction was short-lived. After the first phase, resistance to modernization was small; but most Asian and African leaders distinguished between modernization, which they realized to be necessary, and westernization, which as a form of alienation was to be avoided. Indeed, it might almost be said that the essential problem facing them was how to modernize without westernizing. As one writer has said of Africa, the goal was 'neither the traditional African nor the black European, but the modern African', and this was to be achieved not by resisting and rejecting 'those European elements which modern times demand', but by assimilating and adapting them so that, in combination with elements from the African past, 'a modern viable African culture' would emerge.[1] Behind this, nevertheless, was a lively consciousness of being un-European, an awareness of a cultural inheritance which did not derive from the west and which it was important to retain and integrate into modern life.

It was this sense of difference which underlay the new nationalism of Asia and Africa. Nationalism, it has frequently been argued, was alien to Asian and African societies, 'not part and parcel of an indigenous social system' but 'an exotic institution . . . deliberately im-

1. cf. Legum, op. cit., pp. 102–3.

ported from the west'.[1] Nevertheless we may legitimately question the validity of this generalization. On the whole, it would be nearer the truth to say that any society in the throes of modernization – be it in Europe or be it in Asia – is liable to undergo a process of national concentration. While, therefore, there is no doubt that the nationalist movements of Asia and Africa adapted the techniques and took over the means of expression of the west, it is no less important to emphasize the fact that nationalism itself was 'not born of the revolt against European domination'.[2] This was true in Asia, where its cultural roots reach as far back as in Europe; it was also true in Africa. All the nationalist movements in both continents derived a large part of their driving force from an awareness of a historic past before the European intrusion. This awareness may, like so much western history, incorporate large elements of myth; but the appeal to the ancient African civilizations of the Nile valley, to the chain of states that flourished in the medieval Sudan, to heroic kings like Mansa Musa, the fourteenth-century emperor of Mali, and to outstanding scholars like Ahmad Baba, who taught in the university of Sankore at Timbuktu in the sixteenth century, is a vital element in African nationalism.[3]

It is important to bear in mind the indigenous roots of

1. cf. A. J. Toynbee, *The World and the West* (London, 1953), pp. 70–1.

2. Hall, op. cit., pp. 617–19.

3. cf. Hodgkin, *Nationalism in Colonial Africa*, pp. 173–4. 'I explained', writes Nkrumah in his autobiography (p. 153), 'that long before the slave trade and the imperialistic rivalries in Africa began, the civilizations of the Ghana Empire were in existence. At that time, in the ancient city of Timbuktu, Africans versed in science, arts and learning were having their works translated into Greek and Hebrew and were, at the same time, exchanging teachers with the University of Cordova in Spain. "These were the brains!" I declared proudly. "And today they come and tell us that we cannot do it. . . . But have you forgotten? You have emotions like anybody else; you have feelings like anybody else; you have aspirations like anybody else – and you have visions".'

Asian and African nationalism. The will, the courage, the readiness to undergo persecution – in short, the deep human personal motivation behind the revolt against the west – owed little, if anything, to western example. But will, determination, courage alone were not enough. As the great viceroy, Li Hung-chang, pointed out at the time of the Boxer rebellion, resistance to the west was worse than useless until conditions changed.[1] The history of the twentieth century has been the history of this change in conditions. Its result has been a revolution in the relative position of Asia and Africa in the world which is almost certainly the most significant revolution of our time. The resurgence of Asia and Africa has given a quality to contemporary history different from anything that has gone before; the collapse of empire is one of its themes, but the other, and more significant, is the advance of the peoples of Asia and Africa – and, more slowly but no less surely, of Latin America – to a place of new dignity in the world.

1. cf. Romein, op. cit., p. 8.

# VII

# THE IDEOLOGICAL CHALLENGE

*The Impact of Communist Theory and Soviet Example*

EVER since the Russian revolution of 1917 people have depicted the drama of contemporary history as a tremendous conflict of principles and beliefs, a clash of irreconcilable ideologies. They have compared it to the struggle between medieval Christianity and Islam or between Catholics and Protestants at the time of the Reformation and have seen in it 'the most vital issue of our time', 'the great continuing conflict of the twentieth century'.[1] In reality, the position is a good deal more complicated than such formulations suggest. The lasting significance of the ideological struggle, it is beginning to appear, was to set the stage for more far-reaching changes – the emancipation of the peoples of Asia and Africa, for example – and its relevance to the conditions of the later twentieth century and to such overriding problems as the feeding of a burgeoning world population is increasingly questionable. Furthermore, ideologies are so closely bound up in practice with interests that the part they play in events is extremely difficult to disentangle and assess. To take only the most obvious example, it is evident that the conflict after 1947 between the United States and the Soviet Union was not simply a clash of ideologies but a struggle of competing interests, the origin of which can be traced back many years before the Bolshevik revolution of 1917;[2] indeed, if we pay due attention to the underlying

1. cf. J. L. Talmon, *The Origins of Totalitarian Democracy* (London, 1952), p. 1; D. F. Fleming, *The Cold War and its Origins, 1917–1960* (London, 1961), p. xi.
2. cf. above, p. 110.

geo-political factors, it is hard to escape the conclusion that the forces bringing the two countries into collision as world powers would have operated in much the same way if the Bolshevik revolution had never occurred. On the other hand, it is probably true that fear of communism in the west, though it had existed earlier, was intensified when it came to be identified with the formidable military power attained by Russia in Europe after 1945, and that Soviet fears of the capitalist world were magnified in the same way when the ideological conflict was reinforced by the American monopoly of atomic weapons.

The ideological conflict is neither so distinctive a feature of contemporary history as is often assumed, nor is it always much more than useful propaganda for the pursuit of other objectives. The spread of literacy and the rise in its wake of new methods of mass indoctrination led, without doubt, to a marked increase in the power of propaganda framed on crude ideological lines; but throughout the nineteenth century western Europeans had launched diatribes against the 'Asiatic despotism' of the Tsars no less virulent than those later launched against the communists, and there was no aspect of the hatred of the 'godless Reds' which had not already been expressed a century earlier regarding the French revolutionaries. Nevertheless there is no doubt that the rise of a new ideology, which came after 1917 to be identified with Soviet Russia, and the ensuing conflict between the new ideology and the old, profoundly affected the character of contemporary history. What is misleading is to regard it as the central issue to which all else must be subordinated. Marxism was less the cause than a product of a new world situation. But it was no accident that the period which saw the sudden revolutionary advance of industrial technology, the spread of the new conceptions of the state and its functions, and the rise of mass society, also produced a new social philosophy; and we shall hardly be wrong if we describe the emergence of a new ideology as the last component of the new world situation that was coming into existence during the closing decades of the nineteenth century. It was the final proof that a new period of history was beginning. Just as liberalism had emerged after 1789 as the ideology of the bourgeois revolution and a challenge to autocracy and privilege, so at the beginning of the twentieth century Marxism-Leninism emerged as the ideology of the expected proletarian revolution and a challenge to the dominant liberal values. It was an expression of the new forces which social and economic change had released, a doctrine defined to meet the needs of a new age.

1

I have referred specifically to Marxism-Leninism rather than to Marxism, because it is with Marxism-Leninism, to use the clumsy compound consecrated by communist orthodoxy, that we are here primarily concerned. The new doctrines did not, of course, arise fully fledged; their origins can be traced far back in socialist thought, just as the characteristic doctrines of European nineteenth-century liberalism can be traced back into the Enlightenment and beyond. But the specific forms of Marxism-Leninism were new, and it was from these specific forms, rather than the wider tradition of Marxist socialism, that communism as we know it today descended. The ideas propounded by Marx were compatible with many forms of socialism and susceptible of widely varying interpretations: Lenin's doctrines, on the other hand, were in a very real sense a response to the new conditions arising everywhere at the turn of the nineteenth and twentieth centuries. Or, as Stalin was later to say, Leninism was 'Marxism of the era of imperialism and of the proletarian revolution'.[1]

Much has been written about the relation between

1. Joseph Stalin, *Leninism* (London, 1940), p. 2.

Marxism and Leninism, and it is unnecessary to re-open the discussion here.[1] Those with a taste for historical comparisons may perhaps regard Marxism-Leninism as standing in a similar relation to Marx's writings as Pauline Christianity does to the Christian gospels. The important point is that it was Marxism-Leninism rather than 'pure' Marxism that was the starting-point of modern developments. Between Marx's speculations and the official philosophy of Bolshevism, it has been said,[2] there was 'little in common'.

For this there were specific historical reasons. The first was that Marx, though he opened out 'a magnificent vision',[3] was more concerned with analysing the dialectical forces and inner contradictions that would lead to the supersession of capitalism than with the structure of the society that would succeed it. On the most momentous question of all – the question of leadership in a democratic-socialist society – he had nothing precise to say, and he made no attempt to depict the type of government or organization that would be necessary to carry through a successful communist revolution.[4] Furthermore, the basic doctrines of Marxism – formulated between 1846 and 1867, and for the most part nearer to the former than to the latter date – bore the unmistakable imprint of their time. Marxism itself was 'a philosophy born in the west before the democratic age', and Marx and Engels subsequently admitted that the two pamphlets which contained the essence of their teaching, *The Communist Manifesto* (1848) and *The Address to the Communist League* (1850),

1. J. Plamenatz, *German Marxism and Russian Communism* (London, 1954), is as good an account as any.
2. G. A. Wetter, *Dialectical Materialism* (London, 1958), p. 35.
3. J. L. Talmon, *Political Messianism* (London, 1960), p. 224.
4. Marx's principles, wrote Sir John Maynard, *Russia in Flux* (New York, 1962), pp. 294–5, supplied 'the foundation for a commonwealth aiming at the attainment of socialism; but they leave the whole architectural superstructure to the wisdom and taste of the builders'; cf. also Talmon, op. cit., p. 225.

were written at a time of illusion and coloured by ill-founded hopes.[1]

After 1851 the current moved away from the fervour of the revolutionary era, and Marxism moved with it. It would hardly be unfair to say that, prior to Lenin, Marxism became – in the minds of its day-by-day exponents and to a lesser degree perhaps even in those of Marx and Engels themselves – a doctrine of gradualism, chiefly notable for its hostility to all forms of revolutionary activism. This evolution was partly a result of disappointment at the outcome of the revolutions of 1848 and 1849, but even more a consequence of the rapid improvement in the condition of the working classes, which made gradualism seem the most appropriate tactic. In Russia, where Marxism began to make some impact among the left-wing intelligentsia after the publication of Plekhanov's *Socialism and the Political Struggle* in 1883, its most conspicuous characteristic was opposition to the terrorism of the Populists, and it was tolerated for considerable periods by the government as an antidote to the conspiratorial revolutionaries.[2] In Germany, under the influence of Bernstein, the trend was decidedly in the direction of revisionism. Though it clung in theory to its Marxism and condemned Bernstein's doctrines at the party conventions of 1899 and 1903, the great German Social Democratic party, at that date the only substantial organization in the world which claimed to take its stand on Marx, was in fact turning before the end of the nineteenth century into a machine for the defence and propagation of working-class interests in a capitalist society and also for the evolutionary transformation of that society by parliamentary methods.

It was the first great achievement of Lenin to cut through this evolutionary undergrowth. Marx himself, in

1. Plamenatz, op. cit., pp. 168, 217; cf. Engels's introduction to Marx's *The Class Struggles in France* (London, 1934), pp. 13, 16.
2. cf. Maynard, op. cit., p. 293.

his famous *Critique of the Gotha Programme* of 1875, had attacked the gradualism of the German Social Democrats, insisting that the transition from capitalism could only be achieved by means of the dictatorship of the proletariat; but it was Lenin who worked out the techniques of revolution and thus created out of Marxism a new doctrine for a new age. Indeed, it may be said that, with Lenin, who was born in 1870, a new generation, with new problems and a new outlook, took over. Lenin's first important pamphlet, *What is to be Done?*, which he wrote in 1902, was at once the epilogue to the political philosophy of the previous generation and the prologue to political action in the next. Here and in his two subsequent tracts, *Two Tactics of Social Democracy*, written at the time of the Russian-revolution of 1905, and *Imperialism, the Highest Stage of Capitalism* (1916), are set forth what were henceforward to be the main tenets of revolutionary Bolshevism.

Both as a political theory and as a political movement Bolshevism was the creation of Lenin's genius. What E. H. Carr once wrote of Marx applies even more forcibly to Lenin: he 'introduced into revolutionary theory and practice the order, method and authority, which had hitherto been the prerogative of governments, and thereby laid the foundation of the disciplined revolutionary state'.[1] Lenin's work rested on two propositions to which he returned again and again.[2] The first was that 'without revolutionary theory there can be no revolutionary movement'; the second was that a revolutionary class consciousness, far from being a 'spontaneous' growth, could only come to the mass of the workers 'from without', and that the prerequisite of successful political action was 'a small compact core', a revolutionary *élite* of hardened and disciplined party workers. When, in 1903, Lenin succeeded in getting the dictatorship of the proletariat inscribed on the

1. E. H. Carr, *Michael Bakunin* (London, 1937), p. 440.
2. cf. E. H. Carr, *The Bolshevik Revolution, 1917–1923* (London 1950), p. 16.

programme of the Russian Social Democratic Workers' Party, a new era of politics began. The Bolsheviks were only a faction, a splinter of an already splintered revolutionary movement; at the end of 1904 they numbered scarcely more than three hundred, and it was not until 1912 that they emerged as a separate and independent party.[1] But the decisive step had been taken, the line fixed from which – in spite of divisions in the revolutionary front and depression and disintegration during the reaction after 1905 – Lenin never deviated. 'It is not enough', Lenin was later to write, 'to be a revolutionary and an advocate of socialism in general; it is also necessary to know at every moment how to find the particular link in the chain which must be grasped with all one's strength in order to keep the whole chain in place and prepare to move on resolutely to the next link.'[2] Few men in history have equalled Lenin, none excelled him, in this essential quality.

## 2

It is not necessary for our purposes to linger over the history of the years from 1903 to the Russian revolution of 1917, and from 1917 to 1921, by which time the period of civil war and intervention was at an end and the position of the communist government more or less secure. The questions why it was in industrially backward Russia rather than in Germany, as Marx had almost certainly expected, that the revolution took place, and why it was Bolshevism rather than one of the other forms of Marxism that prevailed in Russia, are of considerable historical interest; but the issues which concern us here are different. We are concerned less with the origins than with the impact of Bolshevism, and from this point of view there are three essential considerations.

1. cf. Maynard, op. cit., pp. 308, 318.
2. E. H. Carr, *The Bolshevik Revolution, 1917–1923*, p. 25.

The first is that Bolshevism, or Leninism, reintroduced – what in the period of revisionism had been largely lacking – an active doctrine of revolution. It threw down an open challenge to the existing social order and attacked liberal democracy root and branch, not merely exposing its shortcomings and pressing for them to be remedied, but rejecting its fundamental principles and ideals. The second is that the establishment of the communist state in Russia brought about the polarization of the world into two ideological camps. So long as communism remained an 'ideal', without material backing, its impact was negligible and the small number of its adherents made it unnecessary for existing governments to take it seriously. Its attachment to the existing Russian state, weakened though that was by defeat and civil war, transformed the situation overnight. Just as the 'ideas of 1789' became potent when they were identified with the power of France, so the association of communism with the Soviet Union transformed it from the doctrine of a small subversive minority into a world movement, backed, as time passed, by an increasingly formidable economic and military power. Lenin himself was quick to seize on the point: now, for the first time, he said in 1919, Bolshevism was 'looked upon as a world force'.[1] The third point, however – and for many people the most difficult and paradoxical – is that, in spite of its identification between 1917 and 1949 with the Soviet Union, Bolshevism was from the start, and never surrendered its claim to be, universal in approach and appeal. At the heart of communism, its driving force both for Marx and also for Lenin, was a deeply ethical concern for social justice, for equality between man and man in the sense of non-discrimination on the grounds of sex, race, colour, and class. Marx and Lenin spoke not for one country against others, but in the name of oppressed groups and classes all over the world, and this universality was beyond doubt a main factor in ensuring their influence.

1. cf. A. J. Mayer, *Political Origins of the New Diplomacy*, p. 390.

This does not mean that the claims of communism as a universal ideology and its role as the official doctrine of the Russian state were easily adjusted. On the contrary, it is a matter of historical record that at many critical moments they were a source of tension and even incompatibility. Hostile commentators have made much of the fact; but in the nature of the case it could hardly have been otherwise. For a generation after 1917 the dissolution of the Soviet state would have spelled the end of communism as an established political force. How, then, could it be denied that the immediate tactical necessity of maintaining the position of the Soviet Union must, in case of conflict, take precedence over the long-term interests of international communism? There is no need to enumerate instances, for they have been mercilessly exposed by anti-communist writers. No example is more notorious than the Nazi-Soviet pact of 1939, but perhaps more symptomatic are the sorry story of the Soviet handling of Chinese communism in the 1920s, the prevarications and reversals which marked relations with the non-Russian nationalities within the Soviet Union after Lenin's death in 1923, and – best known of all – the rigid control exercised over the popular republics of eastern Europe between 1946 and 1956.[1] No sensible person would wish to condone these errors and their consequences. But it is important also to see that they arose from an inescapable dilemma, from which no country of any ideological persuasion can ever be entirely exempt. There is no doubt that, after 1929, the policy of the Communist International (or Comintern) was largely dictated by Russian interests; but Seton-Watson is right when he says that, in founding it in 1919, Lenin had no intention of permanently subordinating the other European communist parties to the Russian party,

---

1. H. Seton-Watson, *The Pattern of Communist Revolution* (London, 1960), has analysed these, and other, episodes; cf. particularly ibid., pp. 85–9, 138–46, 242–4, 248–63.

still less to the Russian state.[1] Opponents of communism frequently assert that its ideology is, in practice, merely a cloak to conceal what would otherwise be exposed as a naked pursuit of power politics. Like most cynical views of politics, this is a simplification.[2] Ideologies do not operate in a void, and the relationship between the ideological and the power factors in any situation is extremely complicated and usually beyond our power to unravel; but it is certain that communism could never have exercised so wide and so powerful an influence if – as is often alleged – it had been nothing more than an ideological underpinning of Russian national interests.

There were, in fact, three fundamental reasons for the impact of Marxism on the ideological plane: first, the impression it gave – whatever objections might be raised on the level of theory – of systematic coherence, self-sufficiency, and comprehensiveness; secondly, its universal applicability, particularly when contrasted with the western argument that certain countries were 'not ripe' for democratic self-government; and, thirdly, its peculiar appropriateness as a response to the newly arising conditions of mass civilization. To these, as the new régime was consolidated, were added two further more practical considerations: the apparent strength and efficiency of communist organization, which made an immense impression on political leaders in Asia and Africa, and the convincing evidence provided by the experience and example of Soviet Russia that it was a doctrine that could be made to work.

The mere fact of the existence in Russia of a new political order, coupled with its undeniable achievements in the economic field and its triumph in the war of 1941–5, was a factor of the utmost importance; every success it

1. ibid., p. 75. As Seton-Watson states, Lenin 'wished Moscow to be the Comintern's centre simply because it could give security as the capital of the only communist-ruled country'.

2. cf. R. N. Carew Hunt, *The Theory and Practice of Communism* (Pelican ed., 1963), pp. 21, 171.

registered seemed to demonstrate the validity of its claim to offer an up-to-date alternative to a capitalist system which, on a Leninist analysis, had entered its 'final stage'. Liberal democracy, on the other hand, was on the defensive for most of the period; it seemed, in the mood of dis-illusionment which prevailed after 1919, to have lost its moral drive and its ability to inspire devotion and self-sacrifice, and, after the slump of 1929, its capacity to regulate its own affairs. Of the two conflicting ideologies – the only two, as Lenin insisted, which were possible at the current stage of world history – communism appeared to point to the future and liberalism to be anchored in the past. Like other great historical movements, Bolshevism owed its success not merely to its own power and the enthusiasm it engendered among its disciples, but also to the inner crumbling of the order against which it was directed.

3

Bolshevism divided the world because it was a revolution-ary creed of universal character. It revived the revolution-ary spirit which had been flagging since 1849, anchored it to what to its adherents seemed to be a compelling logical system, and provided it with new forms of organization. The defeat of the Paris Commune of 1871, the last and perhaps the greatest of the many revolts of the Parisian workers, had marked the end of a period; it proved, as Engels had foretold, that the time was past for re-shaping society 'by simple surprise attack' – by the strategy, that is to say, of 1791 and 1792 – and reinforced his conclusion that a new revolution would only be possible 'in the wake of a new crisis'.[1] Bolshevism, as created by Lenin, provided the strategy for the new crisis brought on by the war of 1914–18.

It was, in its early phase, only one of a number of

1. *The Class Struggles in France*, pp. 13, 21, 25, 135.

revolutionary movements which foreshadowed a new age. In France the lead was given by Georges Sorel, whose *Réflexions sur la violence*, published in 1905, were in some ways an even more drastic rejection of gradualism than Lenin's *What is to be Done?* Sorel preached the inevitability of class war and the need for proletarian revolution to bring a classless society; like Lenin, he advocated 'direct action' under the leadership of an 'audacious minority', and the use of violence to destroy the bourgeois state. Trotsky, also, and Rosa Luxemburg were propounding similar views at the same period, independently of Lenin. The recrudescence of revolutionary philosophies was, indeed, a distinctive feature of the age. Not all were Marxist; some derived from Bakunin, others from Proudhon, others from Lassalle; and some led in the direction not of socialism but of fascism. But all were marked by a reaction against progressive liberalism and a revulsion towards activism in politics. They signified the close of what Marx called the 'long malaise' following the bourgeois revolution, the ending of the 'interlude in the great drama' which the Swiss historian Burckhardt, almost alone among non-socialist thinkers of his generation, had gloomily foretold in 1871.[1] It may be an exaggeration to say, with Halévy, that by 1914 'no responsible statesman ... felt safe against the perils of some kind of revolutionary outburst';[2] but it is certain that, from around 1905, the challenge to liberalism, which is the outstanding feature of contemporary history on the plane of ideas, was in the air. It was Lenin's achievement to bring it down to earth.

The reasons why it was Leninism, or the Leninist form

1. Marx, *The Eighteenth Brumaire of Louis Napoleon* (trans. E. and C. Paul, London, 1926), p. 27 (where 'langer Katzenjammer' is translated 'long fit of the dumps'); J. Burckhardt, *Judgements on History* (London, 1959), p. 209.

2. E. Halévy, *The World of Crisis of 1914–1918* (Oxford, 1930), p. 19.

of Marxism, that finally emerged as the great antagonist of liberalism are many and have been discussed at length. What practically no one would deny is that it would never have occurred without 'the powerful and extraordinary personality of Lenin' himself.[1] Lenin's revolutionary genius is a prime factor which cannot be left out of account. It was his insistence on doctrinal integrity, even at the cost of splitting his party, his ruthless refusal to compromise, his clear grasp of essentials, but above all his unbending revolutionary will that enabled him to forge an instrument capable of taking over power in Russia when the moment came. No one but Lenin would have made the famous reply when, in June 1917, Tsereteli asserted that there was no party in Russia that would dare to assume sole authority: 'Oh yes, there is; our party is prepared at any moment to take over the entire power.'[2] It was due to Lenin personally that Russian socialism was extricated from the labyrinth of speculative reflection which, at the close of the nineteenth century, was para-lysing its capacity for action. As he wrote in 1904, 'in its struggle for power the proletariat has no other weapon but organization'.[3] He castigated the Marxism of the Mensheviks, with its emphasis on the scientific and evolu-tionary aspects of Marx's teaching, as 'bourgeois-intel-lectual individualism'; Bolshevism, as moulded by Lenin, represented 'proletarian organization and discipline'.[4]

Lenin's emphasis on organization and discipline was in part a reflection of his steely determination to carry revolution from theory to practice, in part a result of his realization that under modern conditions, with all the cards stacked on the side of government, there could (as Engels had pointed out) no longer be any question of

1. Wetter, op. cit., p. 111.
2. Christopher Hill, *Lenin and the Russian Revolution* (London, 1947), p. 225; Carr, op. cit., p. 90.
3. Hill, op. cit., p. 49.
4. Carr, op. cit., p. 36.

seizing power 'by simple surprise attack', and in part a response to the specific conditions existing in Tsarist Russia. In Russia, where the government would scarcely tolerate the liberalism of Miliukov and Struve, there was no room for the evolutionary, revisionist Marxism that was gaining ground in the west; 'the nature of the political and social system drove almost all educated Russians into opposition.'[1] It was this fact which explained why Russian socialism drew away from that of the west, and why it was in the Leninist form of Marxism that the revolutionary challenge to liberal ideology came to be anchored.

In Lenin, the Marxism taken over from the west merged with the Russian revolutionary tradition of Chernyshevsky, Tkachev, and Nechaev.[2] But to describe Bolshevism, as is sometimes done, as Russian Marxism is to misunderstand the import as well as the impact of Lenin's revolutionary genius. Lenin worked for and wanted revolution in Russia, but he never conceived of the Russian revolution in isolation or of Marxism as confined to Russia. The doctrine of 'socialism in one country', as propounded by Stalin after 1924, was no part of the Leninist canon.[3] When Lenin arrived in Petrograd from Switzerland in April 1917, he was convinced that the Russian Social Democrats, by seizing power in their country in the name of the workers, would precipitate social revolution in the west and anti-colonial risings in the east. On his analysis of the situation – an analysis that events proved to be wrong – the effect of the great war was to put an intolerable strain on the industrial powers engaged in it, the only outcome of which could be proletarian revolution. Early in 1919 Zinoviev confidently predicted that

1. Seton-Watson, op. cit., p. 12.

2. cf. F. Venturi, *Roots of Revolution* (London, 1960), pp. xi, xiii, xxiv, xxix.

3. For 'socialism in one country', the doctrinal controversies between Stalin and Trotsky and the differing interpretations they placed on Lenin's words, cf. I. Deutscher, *Stalin* (London, 1961), pp. 281–93.

'within a year all Europe will be communist'.[1] It was only when events falsified this prediction that the communist position began to change, and, without throwing over the doctrine of world revolution, Lenin and Stalin after him began of necessity to concentrate on the immediate task of ensuring the security of the Soviet Union in a hostile world.

Through all the contortions of policy which followed, Lenin's original intention was never repudiated, and, indeed, could never be repudiated without treason to the basic concepts of Marxism-Leninism. The object was not to change the social order in one country but to change it everywhere. Social Democracy had foundered on the rock of nationalism, which had destroyed the Second International. Communists on the contrary, owed their first duty, whatever their nationality, not to their nation but to their class. This principle was, of course, often disregarded and sometimes flagrantly transgressed. The longer Stalin remained in power, the more communist policies appeared to take second place to Russian national interests, and few facts perhaps did more to bring the movement into disrepute. Revolutionary movements in foreign countries were staged, or called off, according to whether or not they fitted in with Soviet policy, and almost the whole generation of 'old revolutionaries' was summoned to Moscow and liquidated when Stalin, faced with the growing power of Germany, decided in 1935 to call a halt to revolution in favour of the 'popular front'. But although Stalin made it his first duty to preserve and strengthen the Soviet Union – and it would be difficult to argue in the circumstances that he was wrong – he never ceased to be a disciple of Lenin. The concept of 'peaceful coexistence', as it was to be formulated at a later stage by Khrushchev, still lay in the future. Whatever other interpretations of Marxism might be possible, Lenin's Marxism – which Stalin shared – was postulated upon

1. Plamenatz, op. cit., p. 262.

world revolution and unceasing attack on the capitalist system. 'World imperialism', said Lenin in 1919, 'cannot live side by side with a victorious Soviet revolution' – 'the one or the other will be the victor in the end.'[1]

## 4

The first result of Bolshevism, when it was transformed in 1917 from a doctrine into a political force, was to throw a revolutionary brand into a world from which, until 1914, most men believed the spectre of world revolution had been banished. Lenin, with his usual perspicacity, had seen as early as December 1914 that the European war might well prove to be 'the beginning of a new epoch', and as the war dragged on the same conclusion was reached – though this time not with hope but with foreboding – by men of very different stamp and temperament. By 1917 Rathenau, Czernin and Stresemann all realized that what had begun as a European war was rapidly turning into a world revolution.[2] The course of events in Russia confirmed their diagnosis. Trotsky confidently declared that the war had 'transformed the whole of Europe into a powder magazine of social revolution', and in Germany the Spartacists predicted that 'there would be no world peace except on the ruins of bourgeois society'.[3] These predictions underestimated the powers of resistance of the old order; but it remained true that, once again, for the first time since the cooling of the revolutionary ardour aroused by the French revolution, men were divided by an active revolutionary principle. The emergence of a new world was matched by the emergence of a new ideology.

Scarcely less important was the fact that, for the first time in history, it was an ideology which overstepped all geographical boundaries. Whatever the theory may have been, liberalism in 1917 was still in practice limited to

1. cf. Carew Hunt, op. cit., p. 217.  2. Mayer, op. cit., pp. 24, 31. 3. ibid., p. 32.

Europe and the lands colonized by Europeans. Bolshevism knew no such limits of space and race. Far more than the 'ideas of 1789' it was a world-wide ideology. In this, as in many other respects, it reflected a new world situation. Even before the outbreak of war in 1914 Lenin had turned his attention, with remarkable prescience, to Asia, and at the very beginning of the Bolshevik revolution, in December 1917, he and Stalin issued an appeal to the peoples of the east to overthrow the imperialist 'robbers and enslavers'.[1] It was a significant step in a new direction. Lenin was well aware of the importance of the 'hundreds of millions of Asians' who were on the point of becoming 'active participants in the decisions on the fate of the world'. In one of his latest articles, written in 1923, he proclaimed that 'the outcome of the struggle depends in the last resort on the fact that Russia, India, China, etc., constitute the vast majority of mankind', and at the same period Stalin wrote: 'He who wants the victory of socialism must not forget about the east.'[2] It was necessary, he added, 'to convert the dependent and colonial countries from a reserve of the imperialist bourgeoisie into a reserve of the revolutionary proletariat'.[3]

These statements, at the time they were uttered, may have served a tactical purpose – it was the period when Bolshevism had suffered defeats in Germany and Hungary and been thrown back in Poland – but they were a significant indication of the universal connotations of Bolshevik doctrines. Already in 1920 Lenin had emphasized that 'soviet organization' was a simple idea which could 'be applied not only to proletarian but also to peasant, feudal and semi-feudal relations'. It should not be assumed, he said, that 'the capitalist stage of development' was

1. cf. J. Degras, *Soviet Documents on Foreign Policy*, vol. I (London, 1951), p. 17.
2. W. Z. Laqueur, *Communism and Nationalism in the Middle East* (London, 1957), p. 293; Deutscher, op. cit., p. 209.
3. Seton-Watson, op. cit., p. 127.

'inevitable for the backward nationalities'.[1] Looking back from the present, there are perhaps few of Lenin's remarks more percipient than this. If Russia, contrary to the views of the Mensheviks, could telescope its development, reaching socialism without passing through all the stages of capitalism, why should not other 'backward' peoples follow suit? It was this promise of rapid economic and social advance, more perhaps than anything else, that accounted for the basic differences in the reactions to Russian Marxism in Europe, on the one hand, and in Asia and Africa, on the other. 'Asia', one observer has said, had 'less to lose and apparently more to gain than Europe from the Russian brand of communism.'[2]

When we turn to consider the impact of communist theory and Soviet example, it is, therefore, necessary to look first at Europe and then at Asia and the under-developed world. It goes without saying that so vast and controversial a subject cannot be discussed in anything like the detail it deserves; it must suffice if one or two of the more obvious points are briefly indicated.

5

It is usual to describe the impact of communist theory and Soviet example in the west in almost entirely negative terms. As one commentator, writing in 1954, expressed it, the last twenty-five years – that is to say the period that began with Stalin's assumption of supreme power, the collectivization of agriculture and the first five-year plan – had shown that the workers of the west, whom Stalin hoped would be the staunchest allies of the Soviet Union, had not been much attracted by it; 'the better they have known it, the less they have liked it.'[3] With this judgement few would quarrel, at least as an overall assessment. But it is also easy, taking an overall view, to simplify a complex

1. Hill, op. cit., p. 165.  2. Plamenatz, op. cit., p. 342.
3. ibid., p. 270.

process. There have certainly been periods when communism was a powerful political force in western Europe – in Germany before 1933, for example, when the communist party polled over 5 million votes, or in France and Italy after 1945 – and at these times the possibility of its securing a position of dominance exerted a marked influence over the course of events.

Nor should its initial impact be underestimated. Ray Stannard Baker, one of President Wilson's assistants at the peace conference in 1919, pointed out that the Bolsheviks, 'without ever being represented at Paris at all . . . were powerful elements at every turn', and Lloyd George's famous memorandum of 25 March 1919 was haunted by fear of Bolshevism.[1] Particularly after the communist uprising in Hungary the spectre of revolution spreading from Russia dominated the minds and shaped the decisions of the allied statesmen, and was the main argument for granting lenient terms to Germany. 'We are sitting upon an open powder magazine and some day a spark may ignite it,' wrote Colonel House; and Sir Henry Wilson noted succinctly: 'Our real danger now is not the Boches but Bolshevism.'[2] 'Bolshevik imperialism does not merely menace the states on Russia's borders,' Lloyd George told the allied statesmen: 'it threatens the whole of Asia and is as near to America as it is to France.'[3]

These fears were less exaggerated than they have subsequently been made to appear. It is not difficult, looking backwards, to discover reasons why the revolutionary movements in Germany, Austria, Hungary and other countries of eastern Europe were bound to fail;[4] but

1. R. S. Baker, *Woodrow Wilson and World Settlement*, vol. II (London, 1933), p. 64; D. Lloyd George, *The Truth about the Peace Treaties*, vol. I (London, 1938), pp. 404–16.

2. C. Seymour, *The Intimate Papers of Colonel House*, vol. IV (London, 1928), p. 405; C. E. Callwell, *Field-Marshal Sir Henry Wilson*, vol. II (London, 1927), p. 148.

3. Lloyd George, op. cit., vol. I, p. 412.

4. They are listed by Seton-Watson, op. cit., pp. 53–68.

Lenin's plans for turning 'the imperialistic war' into 'an international civil war' were far from an empty dream. But for western intervention in Russia, which tied down the Bolsheviks at the critical moment, the chances of revolution spreading westward were by no means negligible; and Winston Churchill had solid reasons for claiming that the policy of intervention gained, from the western point of view, 'a breathing space of inestimable importance'.[1]

This breathing space the western leaders used to set up, around the western perimeter of the Soviet Union, a *cordon sanitaire* by which they hoped to contain Bolshevism and immunize central and western Europe. They did not yet, for the most part, regard communism as an internal challenge, requiring positive social measures at home; and so long as Russia was crippled by civil war and economic distress, so long also as the western capitalist economy was functioning with a reasonable measure of efficiency, this negative reaction coped adequately with the situation. After the onset of the slump of 1929 these conditions no longer held. Even if it was only a coincidence that the Soviet leaders appeared, through the first five-year plan, to be 'mastering their fate at precisely the same time as the rest of the world fell a hapless victim to the Great Depression',[2] the contrast made an immense impression. What the working classes in the west observed was that the Soviet Union, which had suffered severe unemployment in the period of the New Economic Policy, was now faced by a shortage of labour – and this at a time when unemployment in the west had reached frightening proportions – and that, whereas industrial production in the principal capitalist countries had declined below the level of 1913, that of Soviet Russia showed an almost four-fold increase. In the circumstances of the day it is not sur-

1. W. S. Churchill, *The World Crisis*, vol. v (London, 1929), p. 276.
2. cf. L. Kochan, *The Making of Modern Russia* (London, 1962), p. 274.

prising that more attention was paid to Soviet achievements than to their cost. For the victims of the Great Depression, and for many others as well, they appeared to demonstrate that communism – whatever cautionary qualifications orthodox economists might introduce – was not merely a revolutionary creed, but an economic system which worked while the machinery of capitalism was creaking at the joints.

The reaction to the Soviet impact thus fell into three fairly well-defined stages. The first, running from 1918 to 1929, was almost entirely negative, much like the reaction of Metternich to the French revolution. It sought to contain Bolshevism by isolating it; its instrument was foreign policy, and on the whole it worked, down to 1929, to the satisfaction of western statesmen. The second stage, running from 1929 to 1941, was also a reaction of fear, but much more positive in content. Its characteristic expressions were Fascism and National Socialism, the basic assumption of which, fostered on a grand scale by the depression of 1929, was that liberal capitalism was incapable of withstanding the communist challenge. National Socialism set out to rally the elements in capitalist society – above all the petty bourgeoisie – which felt themselves most immediately threatened. The moral fervour which both Mussolini and Hitler sought to inspire among their followers was whipped up as an antidote to the fervour of Bolshevism, and many of the methods of Bolshevism were invoked in the attempt to generate it. This was the physiognomy which Fascism showed to the world after 1929, and which secured it the toleration, if not the sympathy, of influential elements in non-Fascist capitalist society.[1] The third stage, though its beginnings may be discerned earlier – for example, in the 'new deal' in the United States – only reached full importance after the war

1. The most recent analysis of the character of Fascism, and of the phases of its development, is in E. Nolte, *Three Faces of Fascism* (London, 1965).

of 1941–5. Its basis was the realization that, if the attractions of Marxism were to be combated, it was necessary to show that liberal society could match its achievements, above all by providing security and a higher standard of living for the workers. If it is true that communism 'was not destined to win dominance' in western Europe, it was not because 'the old liberal traditions of Europe' had resumed 'their evolutionary growth' after the middle of the nineteenth century[1] – on the contrary, it would be more correct to say that by the beginning of the twentieth century liberalism was 'a jaded force' compared with what it had been in earlier years[2] – but it was rather the consequence of the deliberate adoption, for the most part in the very recent past, of new social and economic policies.

This is not the place to discuss the character of these social and economic policies, or the merits and demerits of the 'welfare state' or the 'affluent society' to which they are generally agreed to have led. It is possible to argue that the transition from liberal democracy and *laissez-faire* capitalism to the welfare state would have occurred without the impact of Soviet example and the fear of communist contagion; it is possible to hold that it was a response, which would have taken place in any event, to the economic crisis of 1929 and the acceptance of Keynesian economics. Such arguments would be difficult to sustain. The Soviet demonstration that there is an answer to the endemic problems of capitalism, which reached their peak in the crisis of 1929, was not the only factor in bringing about radical changes in the structure of western society by comparison with 1914; but it was certainly an important one. In particular, the whole concept of a planned economy owed much to Soviet example. As Trotsky pointed out, it was the Soviet system which first brought 'aim and

1. Talmon, *Political Messianism*, p. 512.
2. cf. Irene Collins, 'Liberalism in Nineteenth-century Europe', in *From Metternich to Hitler*, ed. W. N. Medlicott (London, 1963), p. 44.

plan into the very basis of society',[1] and its success in eliminating the worst curse of capitalism – namely, unemployment – made it imperative for non-communist governments to turn to planning also. As Mr E. H. Carr has said, 'if we are all planners now, this is largely the result, conscious or unconscious, of the impact of Soviet practice and Soviet achievement.'[2]

6

When we turn from Europe to Asia, the impact of Soviet example and communist theory is much more direct. The launching of the first five-year plan in the Soviet Union in 1928 has been described as the decisive turning-point in the assault on the established position of the European powers in Asia.[3] Certainly, resistance to communism was never so strong in Asia as in Europe and the west. So long as the 'welfare state' continues to function reasonably efficiently, it will be hard to convince the working classes in the west that they have more to gain from communism than they have to lose; their standard of living is higher, their lives are more comfortable, their liberties more congenial, than anything they can imagine under a communist régime. In Asia and Africa these obstacles do not exist, or, at least, do not exist on anything like the same scale. From the beginning those who could expect to gain from communism were far more numerous; the opposing interests were more narrowly based and had been discredited either as authoritarian oligarchies or as the allies of colonial interests, or as both. Western intervention had broken down the barrier of the traditional class structures, but had failed to set up new interests extensive and stable enough to withstand revolutionary pressures.

Two main factors account for the relative strength of the

1. cf. E. H. Carr, *The Soviet Impact on the Western World* (London, 1946), p. 44.
2. ibid., p. 20.
3. T. Mende, *La révolte de l'Asie* (Paris, 1951), p. 10.

communist impact in Asia. The one is that, as a creed, Marxism was 'in some ways remarkably adapted to the needs' of the underdeveloped peoples.[1] The other is that, by comparison with other European nations – the British, the French, the Dutch, the Portuguese, the Belgians – Soviet Russia succeeded to a considerable extent in escaping the stigma of colonialism. This is not to say that, in the Asian republics – in Kazakhstan, for example, or in Uzbekistan – the Soviet Union avoided the problems of nationalism or the anti-colonial reactions which confronted the other European powers. But it showed unusual flexibility in coping with them.[2] The enlightened nationalities policy announced in the early days of the revolution was not followed consistently; in any case it was bound to come up against obstacles when it was translated into practice. But its immediate impact was considerable. What the Soviet Union demonstrated was that the problem of nationalities was 'soluble on a plane of economic equality'.[3] Even before the revolution of 1917 the Russians' unusual understanding of Asian problems and attitudes was widely commented upon; after the revolution there was the same 'perspicacity, originality and imagination'.[4]

'The communists', it has been said,[5] 'have one great virtue in Asia: they are not afraid of simple and drastic action on a gigantic scale.' So far as it goes this judgement

1. Plamenatz, op. cit., p. 339.

2. The question of Soviet 'nationalities policy' is beset by controversy. On the whole, the most objective short account is in G. Wheeler, *Racial Problems in Soviet Muslim Asia* (London, 1962). There is a full, but at some points rather uncritical, account of its origins and early phases in Carr, *The Bolshevik Revolution*, vol. I, pp. 253–380, and R. Schlesinger, *The Nationalities Problem and Soviet Administration* (London, 1956) has published a selection of documents dealing with subsequent developments; cf. also K. Stahl, *British and Soviet Colonial Systems* (London, 1951).

3. cf. H. J. Laski, *Reflections on the Revolution of our Time* (London, 1943), p. 209.

4. cf. Wheeler, op. cit., p. 56.

5. Plamenatz, op. cit., p. 338.

is sound enough. To the sophisticated economies of the west drastic measures on a gigantic scale might do irremediable harm, but in Asia they were capable of bringing immediate benefit to millions of people. One of the outstanding attractions of communism in Asian and African eyes was that it offered the underdeveloped peoples a blueprint for development. 'Capitalism', Kwame Nkrumah once said, 'is too complicated a system for a newly independent nation.'[1] In spite of the 'vast miscalculations' which have occurred both in Soviet and in Chinese planning,[2] most of the leaders in the underdeveloped countries would endorse that view. They would agree that, in Asian and African conditions, the 'gradualist approach' associated with free enterprise 'is almost certain to be self-defeating'.[3] If the mass of the people were to be lifted out of squalor, if hardly won independence were to be preserved, what the west did in several centuries had to be done in Asia in two or three generations. The impact of the Soviet Union was due first and foremost to the practical evidence it provided that this could be done. It was frequently argued that a 'free economy' could achieve as much and more, 'in time'[4]; but time was precisely what was lacking. And if the appalling human cost of planning on the Soviet or Chinese scale was pointed out, the answer was that – in the conditions obtaining over most of Asia, and probably in Latin America and Africa also – the human cost of large-scale planning was unlikely to be greater than the cost of not planning at all. For peoples who had known little of the traditional western freedoms – and that is the case, for example, of the fellaheen of Egypt or Iraq and the labourer in the paddy-fields of

1. Kwame Nkrumah, *Autobiography* (Edinburgh, 1959), p. vii.
2. cf. A. Nove, *The Soviet Economy* (London, 1961), p. 294.
3. cf. B. H. Higgins, *Economic Development. Principles, Problems and Policies* (New York, 1959), p. 454.
4. cf. R. Harris, *Independence and After. Revolution in Underdeveloped Countries* (London, 1962), p. 45.

Burma – the consequential restrictions and compulsions were a small price to pay.

It would nevertheless be a mistake to place too much emphasis on the economic aspects of the Soviet impact on Asia. As Isaac Deutscher has pointed out,[1] it is in the fields of social policy and education – not in wealth and productivity where it can more than hold its own – that the west has found it most difficult to match the Soviet advance. And Walter Laqueur has insisted that 'the religious and ethical elements in communism have been of far greater importance' than the economic ones.[2] It could hardly escape the notice of Asian and African leaders, for example, that the Russians did more in a quarter of a century for the education of the peoples living in the Arctic circle and in the Caucasus, who in 1917 had not even a written language, than the British did in India in an occupation of nearly two hundred years. It would be foolish also to underestimate the political attraction of communism for the lawyers, the scientists, doctors, technologists, and managers who – in association with army officers from similar social strata – were emerging as the dominant element in Asian and African societies. To them it offered prospects of leadership and genuine achievement, and what they may have had to give up as individuals – in Asian society it would usually not be much – they stood to gain in professional standing.[3] Communist forms of political organization have marked affinities with the traditional Asian system of an authoritarian state

1. cf. I. Deutscher, *The Great Contest. Russia and the West* (London, 1960), p. 78.

2. Laqueur, op. cit., p. 284.

3. As Laqueur expressed it (ibid., p. 273): 'They are to be the masters, teachers, builders, makers of the new country, and the new men; they will be lavishly equipped with all facilities that can promote their work; rather than foreign bodies in their old communities they will be the centres around which a new community crystallizes; the more comprehensive the new structure grows, the higher will be their place on the pyramid of functions that they themselves have to organize.'

which is the incarnation of absolute law.[1] On the other hand, civil and political liberties of the western type carry less weight than we are apt to think in societies where it has always been regarded as natural for governments to impose duties and obligations, rather than protect and safeguard individual rights. Moreover, it cannot be assumed that democratic institutions of the western type will necessarily be effective under Asian conditions.[2] In countries where the contrast between wealth and poverty is still extreme, and where parliamentary institutions can readily be manipulated in the interests of the former, dictatorship may be the only method – or at least the only practical method immediately available – of securing democracy in the original sense of the word, as used by Aristotle: that is, as the antithesis of aristocracy or plutocracy, or of the predominance of any other narrow class interest exercising power on the basis of the control of property. Asian democracy in practice is apt to conform to Stalin's description of democracy in capitalist countries – 'democracy for the strong, democracy for the propertied classes'.[3]

There is no need, in any of this, to idealize Soviet society or to minimize either its harshness to minorities or its inefficiency and waste. We are concerned simply with an historical situation; and it is part of that situation that a system derived from Marx and Lenin appeared to many of those concerned to be better suited to Asian conditions than any practicable alternative. It did not follow that it must be the Soviet or Russian system; indeed, the evidence would indicate that the adoption of a system on the Russian model is no longer very likely. Since the establishment of the Chinese People's Republic in 1949 no

1. cf. Mende, op. cit., p. 93. On the other hand, Harris (op. cit., pp. 7, 11) emphasizes the difference between the authoritarianism of east Asia and the situation in south Asia where 'there are no strong traditional barriers to the growth of democracy'.

2. cf. Mende, op. cit., p. 14.

3. cf. Carr, *The Soviet Impact on the Western World*, p. 11.

communist party has won control in any country in Asia, or in Africa or Latin America.[1] That does no mean, however, that Marxism as interpreted by Lenin or by Mao Tsetung has lost its intellectual attraction. Except for India, where nationalism had made substantial progress before the Russian revolution of 1917, most of the nationalist movements in Asia had a strong Marxist component from the beginning, and the ideological force of Marxism remained strong even for leaders who, like Nehru, rejected communism as a political system. Hence it would be wrong to measure the strength of Marxism as an ideology by the success or lack of success of the Asian communist parties. More important in the long run was the fact that the missionary role which was filled after the First World War by American democracy under the inspiration of President Wilson, and which mainly affected Europe, was filled after the Second World War by Soviet democracy, and mainly affected Asia. It did so for two reasons. First, its content was primarily social, and thus it corresponded to the aspirations awakened throughout Asia for social reform, whereas the content of western democracy was largely political. Secondly, unlike western democracy, which appealed mainly to the middle classes, it was able to communicate with all levels of society and offer them a new sense of solidarity with a place for all in the system. When Lenin said that 'politics begin where the masses are' – 'not where there are thousands, but where there are millions, that is where serious politics begin'[2] – he was speaking of Russia, not of Asia; but it was in Asia, with its teeming millions, that his saying bore fruit. Communism offered a new principle of order to societies which western intervention had thrown into the melting pot.

1. North Korea and North Vietnam are not exceptions, since the agreements of 1953 and 1954 only recognized a *status quo* which had existed before the outbreak of war.

2. Lenin's statement was made at the seventh congress of the Russian communist party on 7 March 1918; cf. V. I. Lenin, *Selected Works*, vol. VII (London, 1937), p. 295.

Its bold solutions, its willingness to cut through tangles, above all its dynamic belief in itself and its mission, raised it, for Asian purposes, above the cautious pragmatism, linked with a crippling respect for entrenched interests, which seemed to mark the western approach to Asian problems.

## 7

It is only necessary to compare the world situation in 1900 with that sixty years later to see how profoundly in the interval the impact of the new ideology had changed the balance. Where, at the beginning of the century, the liberal democratic order, anchored to a *laissez-faire* economic system, appeared to be advancing unchallenged, by 1960 the world was divided. A third of the inhabitants of the globe were outside capitalist society and integrated in a rival system where overall social and economic planning was the rule and production was no longer regulated by the profit motive. This was the widest consequence of the impact of Marxism-Leninism. Belief in the inexorable laws of capitalist economics had been broken, and even in the west the concept of a 'free' economy had largely given way to the prevalence of 'mixed' economies with some degree of planning at the top, an expanding 'public sector' and a measure of government regulation that would have been unthinkable sixty years earlier.

There is no doubt that these developments have changed the character of the ideological conflict which was so powerful between 1917 and 1956, so much so that it has become common to speak of 'the end of ideology'[1] and to predict that 'one day', perhaps not far distant, the two systems "will meet in the middle".[2] So far as the Soviet Union is concerned, this argument is plausible enough. It is not only that western society has liberated itself from

1. cf. D. Bell, *The End of Ideology* (paperback ed., 1965), pp. 402–4.
2. cf. Nove, op. cit., p. 303.

the extremes of *laissez-faire* capitalism; Soviet society, also, has entered a period of rapid change. Once the phase of 'primitive socialist accumulation' in the Soviet Union had ended, the transition from a state of scarcity to a state of plenty quickly generated significant social and political developments. Already under Stalin a managerial technocracy had come into existence, similar in many ways to the managerial stratum which emerged in the west after the development of industry had removed ownership and active control from the hands of the capitalist entrepreneur and transferred it to an anonymous and amorphous body of stock- and shareholders. Under Khrushchev conservative elements gained further ground and the revolutionary fervour of the early Bolshevik generation became a thing of the past. As in the west, the mass of the people were more interested, by the close of the sixth decade of the twentieth century, in enjoying the benefits of affluence than in prosecuting an ideological crusade. These were significant facts. They indicated – in conjunction with such developments as peaceful coexistence and thermonuclear stalemate – that the 'cold war' which had been the mark of the transitional period, was drawing to a close. It is nevertheless important not to misunderstand their significance. As Schumpeter once said, 'to confuse the Russian with the socialist issue' is 'to misconceive the social situation of the world'.[1] Even if the Soviet Union is in process of becoming a conservative society – much as France became a conservative society after the basic aims of the French Revolution had been achieved – over much of the rest of the world the problems raised by Marx and Lenin remain unsolved and the appeal of their doctrines to the underdeveloped peoples is still powerful.

The decade following the Twentieth Soviet Party Congress in 1956 nevertheless saw two important developments. The one was the gradual subsidence of the ideo-

1. cf. J. A. Schumpeter, *Capitalism, Socialism and Democracy* (London, 1961), p. 405.

logical conflict between the Soviet Union and the United States and the indications that, confronted by a new constellation of forces, the two countries were seeking to find a basis for *rapprochement*. The other was the simultaneous emancipation of Marxist ideology from Russian tutelage. With the emergence of 'national communism' in eastern Europe and elsewhere, with the recognition of the possibility of 'separate roads to socialism', and, above all, as a result of the enhanced status of communist China, Marxism and Leninism ceased to have even the appearance of specifically Russian doctrines. This change did not, of course, alter the fact that Soviet communism was still one of the most powerful forces in the world. But it is also true that the impact of Marxism became broader, more varied and less monolithic than in the days of Stalin. This was apparent not only within the communist *bloc*, where Mao Tse-tung provided an alternative version of Leninism to that propounded by Moscow,[1] but also in non-communist Asia and Africa. In Indonesia, for example, although the Indonesian communist party played an important role until its bloody suppression in 1965, the characteristic ideology was Marhaenism, the Indonesian interpretation of Marxism.[2] And it was Nehru's conviction that, for India also, there was 'only one solution – the establishment of a socialist order . . . with a controlled production and distribution of wealth for the public good'.[3] This did not imply rigid copying of Soviet political or economic methods, still less alignment with the communist *bloc*; but it did imply the adoption of the broad criteria of Marxist thought.

It is not, of course, necessary to assume that such a

1. cf. S. R. Schram, *The Political Thought of Mao Tse-tung* (New York, 1963), pp. 56 f., 78 ff.

2. cf. J. Mintz, *Mohammed, Marx and Marhaen* (London, 1965).

3. Jawaharlal Nehru, *An Autobiography* (London, 1936), p. 523; cf. also K. T. Narasimha Char, *The Quintessence of Nehru* (London, 1961), pp. 140–5, where further statements of a similar character are assembled.

solution will be achieved – 'the socialist Phoenix may fail to rise from the ashes'[1] – but so long as it is pursued it is unlikely that Marxist ideology will lose its force. The effects of Russian experience in this respect have been double-edged. On the one side, the attraction of Marxism-Leninism was heightened by the demonstration in the Soviet Union of its capacity to transform the living conditions of a backward society; on the other, the leaders in many Asian and African countries were repelled by the way it was manipulated in Russia under Stalin. But the emotional and intellectual appeal of Marxism only depends to a limited degree on Russian experience and Russian example. More fundamentally, its emergence as one of the leading ideologies of a new age was a reflection of the conviction that liberal capitalism was unable to solve the problems of modern society, and until the falsity of this belief has been demonstrated on a world-wide scale – which is still not the case – the impact of Marxism as a world force is unlikely to diminish, though its forms may change.

It may be true, so far as the industrialized countries of the west are concerned, that developments since 1945 have demonstrated the ability of capitalist society to come to terms with the conditions of the modern world. Though persistent inflation, 'high-level underdevelopment' and 'partial technological stagnation'[3] have bred a certain scepticism, few people would deny that Keynesian economics, the maintenance of full employment, social services, and the redistribution of income by taxation have restored the stability of the private-enterprise system

1. cf. Schumpeter, op. cit., p. 57.
2. The classical analysis of these problems is, of course, J. K. Galbraith's *The Affluent Society* (London, 1958), from which a whole category of literature has descended. Schumpeter also was sceptical of the ability of neo-capitalism to 'survive indefinitely' (op. cit., p. 419): cf. also J. Robinson, *Economic Philosophy* (London, 1962), particularly p. 116 for 'high-level underdevelopment' in the United States.

which appeared before 1939 to be on the verge of collapse. When we turn to the underdeveloped world, the position is entirely different. It is not, as is often said, that under the adverse conditions in Asia, Africa, and Latin America capitalism based on the profit motive will not work, but rather that the better it works and the more efficient it becomes as an economic system, the more likely it is to increase social disequilibrium and give rise to revolutionary social tension. But more important is the fact, of which Gunnar Myrdal has done so much to make us aware,[1] that the effect of the high standards of living achieved within the affluent societies of the west has been to accentuate, rather than to mitigate, the long-standing inequalities in the world distribution of goods and services. Whereas in 1945 the income of the average citizen of the United States was twenty times that of the average Indian, by 1960 it was some forty times.[2] In spite of aid, loans and technical assistance, in other words, the gap between the industrialized and the underdeveloped peoples is widening, and there is no evidence that such measures as have been taken have remedied this situation, or are capable of remedying it. In 1957 the share in world trade of the non-industrial nations, which contained over three-quarters of the world's population, was lower than in 1928, when they were still undeveloped, and the average standard of living of humanity as a whole is still below the level of 1900.[3] Everywhere, in short, the 'law of cumulative inequality' is at work, creating discrepancies more glaring than in the past.[4]

The problems which these inequalities pose are deeply rooted in the existing economic system, and it would be unrealistic to suppose that there is any simple solution to

1. cf. G. Myrdal, *Beyond the Welfare State* (London, 1960), pp. 119 ff., 164–5.

2. cf. Evan Luard, *Nationality and Wealth* (London, 1964), p. 322.

3. cf. G. Myrdal, *An International Economy. Problems and Prospects* (London, 1956), pp. 2, 149.

4. Luard, op. cit., p. 216.

them. But they are also the operative reason why Marxism-Leninism remains an active force in the world today. To regard it merely as an ideological weapon of the Soviet government would be to misunderstand its historical role. The fact that it was closely identified, for thirty or forty years after 1917, with the Soviet Union was a consequence of historical circumstances which were of immense importance at the time, but which no longer prevail. Lenin himself pointed out that, once the proletarian revolution had achieved a measure of success, Russia would 'cease to be the model country',[1] and there are many indications that, as it evolves and is adapted to other circumstances in other parts of the world, Marxism is beginning to modify or cast off the specifically Russian features it acquired between 1928 and 1953. To say this is not, of course, to make the mistake of underestimating the part played by the Soviet Union in recent history. But the significance of Marxism transcends its importance as the ideology of the Soviet state. Historically, Marxism as interpreted by Lenin and Mao Tse-tung is significant because it provided an alternative for the emergent peoples, to whose conditions the liberal economic system of the west and the political and social institutions associated with it were not easily adaptable. It was not the only conceivable alternative system; but it was the only one which had the dynamism the comprehensiveness and the emotional appeal which their situation demanded. If we are to measure its impact we must see it not simply as a Soviet Russian ideology, but – as Lenin saw it – as a universal force with a universal mission. It has already shaped twentieth-century society on lines different from anything known in the past; and its force is not yet spent.

1. *Sochineniya*, vol. XXXI, p. 1.

# VIII

# ART AND LITERATURE IN THE CONTEMPORARY WORLD

## The Change in Human Attitudes

IF, as it has been the purpose of this book to show, contemporary history is different in most of its basic preconditions from what we call 'modern' history, if the contemporary period marks the onset of a new epoch in the history of mankind, it would be reasonable to expect to see this change mirrored not only in the social environment and in the political structure, but also in human attitudes. It is, of course, true – as Marx was always careful to emphasize – that the relationship between the social infrastructure and the superstructure of 'sentiments, illusions, habits of thought and conceptions of life' raised upon it is extremely complex;[1] and we should be foolish to expect any direct coordination between them. But it would also be surprising if the temper of literature and other forms of human self-expression had not been affected by the new social order produced by a new technological civilization. What I shall attempt to do in conclusion, therefore, is to examine some of the more outstanding changes in human attitudes in the last three-quarters of a century and see how far they indicate the emergence of a new outlook on

1. cf. Karl Marx, *The Eighteenth Brumaire*, trans. E. and C. Paul (London, 1926), p. 55. The importance of distinguishing between 'the material transformation of the economic conditions of production' and 'the legal, political, religious, aesthetic or philosophic' forms was also emphasized by Marx in a famous passage in the preface to *The Critique of Political Economy* (ed. 1904), p. 12. As is well known, Engels also protested after Marx's death against the accusation that he and Marx had maintained that the ideological superstructure responded directly and unconditionally to economic conditions; cf. G. A. Wetter, *Dialectical Materialism* (London, 1950), pp. 38–40, 50.

the world and a new approach to basic human problems.

Our starting-point will be the 'disintegration of the bourgeois synthesis'[1] which stares us in the face as the nineteenth century draws to a close; the central point of our inquiry will be whether any new synthesis has yet succeeded it, or whether we can at least discern the elements of a new synthesis. Two points, in particular, will call for attention: the one is the degree to which our attitudes have been reshaped by the revolution in science and the impact of technology, the other is how far the new mass society of our times has arrived at distinctive forms of expression of its own. These are questions which have been hotly debated, often in terms of value-judgements which are subjective and largely irrelevant. We have had a surfeit of moralizing on the decadence of modern art and music, the gulf which allegedly separates them from everyday life, the spiritual erosion of western civilization, and, more recently, the shortcomings of the resurgent peoples of Asia and Africa. Such judgements cloud rather than clarify the issues, and I shall try to avoid them. The pessimism which sees all change as change for the worse is a recurrent theme of history, which history has recurrently refuted. But when all this has been said, a real question remains, which the historian cannot simply ignore on grounds of lack of technical qualifications.

In following the process of change as it affects human attitudes, we can fairly easily distinguish three main phases or periods. The first, which extends from about 1880 to the First World War, was marked above all by reaction against the tradition of the past four hundred years; the second, roughly equivalent to the inter-war years, but extending back to the decade before 1914, was a period of great experimentation in new modes of expression; in the third, which followed the Second World

---

1. cf. G. Bruun, *Nineteenth-Century European Civilization, 1815-1914* (London, 1959), p. 186.

War, much of the experimentation of the inter-war period was left behind, but it was still not easy to perceive the crystallization of a new outlook on the world. This should not be surprising. When we consider the extent of the upheaval of the last half-century and the magnitude of the adjustments to be made, it would be unrealistic to expect the rapid emergence of a new unifying culture. In other respects – for example, in the shaping of new political terms of reference – we can say with some confidence that the transition from one era to another has been completed. In respect of our basic human attitudes we must expect slower progress. The diffusion of a new cultural pattern requires a period of stability such as we have not experienced since 1914, but which may now be beginning. Even then, there is the question whether the old liberal synthesis, which was the mark of the nineteenth century, will be succeeded by anything comparable in scope and influence.

1

For the historian it is easier to trace the disintegration of old attitudes and patterns than the formation of new ones. The central fact marking a break between two periods was the collapse – except in formal education, which was thereby increasingly cut off from the mainstream of social development — of the humanist tradition which had dominated European thought since the Renaissance. The attack on humanism took many forms and came from many directions; but at its heart was disillusion with humanism itself, and it was the discrepancy between its professions – namely, respect for the dignity and value of the individual – and its practice – namely, the dehumanization and depersonalization of the working classes – that initiated the revolt. What brought it to a head, after a period of growing disquiet, was the sharp deterioration of conditions in town and factory resulting from the new industrialism,[1] and it was fostered by the new preoccupation with the maladies of poverty, unemployment and distress which marked the generation from Henry George's *Progress and Poverty* (1879) to William Beveridge's *Unemployment* (1909). It found its most eloquent expression in the best of Zola's writing, notably in his greatest novel *Germinal* (1885), with its insistent hammering on the themes of hardship, endurance, darkness, mass action and mass suffering. Something of the same quality infused Gerhart Hauptmann's greatest drama, *The Weavers* (1892).

Works such as *Germinal* exposed the hollowness of humanist professions, the implicit contradiction at the heart of liberal philosophy between human dignity and equality in theory and economic inequality and indignity in fact. At the same time Nietzsche – the mature Nietzsche of *Also sprach Zarathustra* (1883–5) and *Beyond Good and Evil* (1885–6) – was savagely attacking its moral pretensions, tearing away the ideological veil erected to conceal the power structure on which the social order was based, and hammering home the brutal truism of the will to power. 'Seek ye a name for this world? A solution for all its puzzles? . . . This world is the will to power – and nothing else.'[2] With a directness without parallel before him Nietzsche penetrated through the optimism of his day, the facile belief in progress automatically assured by natural selection and the survival of the fittest, the assumption that man, the individual, is an infinite reservoir of possibilities and that all that is necessary is to rearrange society for these possibilities to prevail. Morality was 'itself a form of immorality';[3] philosophy from Plato to Hegel had falsified reality and degraded life. 'Nothing has been bought more dearly,' Nietzsche proclaimed, 'than

1. For this deterioration, cf. above, pp. 50–52.
2. Friedrich Nietzsche, *Werke* (ed. K. Schlechta), vol. III (2nd ed., Munich, 1960), p. 917.
3. ibid., p. 527.

that little bit of human reason and sense of freedom which is now the basis of our pride.'[1] It was this frontal attack on the values and assumptions on which all western culture was based that made him, after 1890, the inspired prophet of the new generation in Europe.

Nietzsche's disruptive influence on the nineteenth century's picture of intellectual man, the purposive master of his own fate, was reinforced by the work of the French philosopher, Henri Bergson, with his assertion of the superiority of intuition over intelligence. It was reinforced also by new trends in the physical sciences and by the impact of new psychological insights. Both contributed, with increasing force as time passed, to the decline of the certitudes which had sustained the commonly accepted picture of man and the universe. Science, in the first place, dissolved the old concept both of nature and of man's place in nature. The French mathematician, Henri Poincaré, denied that science could ever know anything of reality; all it could do, he asserted, was to determine the relationship between things. In England a similar view of the world as a structure of emergent relationships was put forward in F. H. Bradley's *Appearance and Reality* (1893) and developed by Whitehead and the relativists. 'Nature by itself', Bradley maintained, 'has no reality'; the idea that nature was 'made up of solid matter interspaced with an absolute void', which had been inherited from Greek metaphysics, was untenable and must be discarded.[2] Space, Bradley asserted, was only 'a relation between terms which can never be found'.[3] Thus nature, which from the time of Giordano Bruno had been a fixed point of reference – the totality of things and events which man encountered around and about him – began to retreat into inaccessibility; it became an intricate network of relations and

1. cf. Hans Kohn, *The Mind of Germany* (London, 1961), p. 214.
2. cf. F. H. Bradley, *Appearance and Reality* (2nd ed., London, 1902), pp. 288, 293.
3. ibid., p. 38.

functions which was beyond common experience and could only be conceptualized abstractly, until finally it dissolved into a 'lost world of symbols'.[1]

The trend of modern science was to suggest that the universe is unintelligible, senseless and accidental, and that man, in Eddington's phrase, is 'no more than a fortuitous concourse of atoms'.[2] Such views, as they passed into wider circulation, could not but have a dissolving effect, and the same was true of the new psychology of Pavlov and Freud. Freud, whose *Interpretation of Dreams* appeared in 1899, must be ranked with Lenin as the herald of a new age. Although his main influence was not felt until after 1917, he was a figure of formidable stature and influence, with whom, in the scientific field, only Einstein could compare. The Freudian theory of the subconscious had an immeasurable impact, above all by destroying the image of man as a coordinated individual responding intelligently and predictably to events. Freud's discovery that man's actions could be motivated by forces of which he knew nothing exploded the individual's illusion of autonomy, and sociology which, in Dewey's phrase, conceived 'individual mind as a function of social life', worked in the same direction. If science left man groping after an elusive external reality, Freud left him seeking in vain for the reality of his inner self.

2

The effects of these revolutionary changes in outlook on all fields of literary and artistic expression are too obvious to need more than illustration. We know that writers like Henry James and Virginia Woolf quickly took note of the new psychological insights; we know that the early cubists,

1. cf. R. Guardini, *The End of the Modern World* (London, 1957), p. 91.

2. A. S. Eddington, *The Nature of the Physical World* (Cambridge, 1928), p. 251.

while they were living in the Bateau Lavoir on the slopes of Montmartre, learnt about the new scientific outlook from the amateur mathematician, Princet; we know that Eliot, as a student at Harvard, studied and wrote on Bradley. Such cases are, however, rare and it would be misleading to stress them. Changes in literature and art and changes in philosophy and science occurred simultaneously and largely independently, and the effect of the latter was to play upon and accentuate a process of disintegration in the former which was already under way. Confidence in the power of art to reflect the true nature of reality was already withering, and science only confirmed the existing awareness that truth does not, after all, conform to instinctive feelings or immediate perceptions. 'I arrange facts in such a way', wrote Gide in 1895, 'as to make them conform to truth more closely than they do in reality.'[1] 'Art does not reproduce what can be seen,' said Paul Klee; 'it makes things visible.'[2]

From 1874, when the first impressionist exhibition was held in Paris, it was impossible to overlook the disintegration of the artistic tradition that had held sway since the Renaissance. It is true that the impressionists – Monet, Pissarro, Renoir, Degas, Sisley – were not revolutionaries breaking with the past; as Gauguin said, they 'kept the shackles of representation' and remained within the tradition of realism, but now it was 'a realism evaporating into the immaterial reality of air and light'.[3] It was when the problem arose of filling the void which the dissolving effects of impressionism left behind that the need for new standards and for a break with the past became explicit, and almost simultaneously painting, writing and music

---

1. André Gide, *Paludes* (ed. Paris, 1926), p. 21.
2. Klee's essay of 1920, of which these are the opening words, is printed in translation by W. Grohmann, *Paul Klee* (London, 1958), p. 97; cf. also G. di San Lazzaro, *Klee: A Study of his Life and Work* (London, 1957), p. 105.
3. cf. *New Cambridge Modern History*, vol. XI, p. 158.

set off on deliberately revolutionary lines. There can be no question here of following the process step by step, from van Gogh, Cézanne and Gauguin to Kandinsky, Picasso and Jackson Pollock, from Debussy to Schönberg, Bartok and Webern, from Mallarmé and Rimbaud to Éluard and Pound. All we can do is to pick out a few trends, of which the most obvious was the rejection of accepted artistic forms.

The preoccupation with form was characteristic, for it was the bankruptcy of the old forms and the need for new ways of coming to terms with a new type of human being, the new emotions which filled him, his new relationship with the world about him, which dominated artistic expression. It led inexorably from symbolism to expressionism and cubism, and behind it was a repudiation of the preoccupation with nature – with three-dimensional space, scientific perspective, *sfumato* and *chiaroscuro* – which had dominated European art since the Renaissance. Whistler's Nature is usually wrong!' stated the challenge of a new generation to the whole existing tradition. The central purpose of the cubists and the German expressionists was to get away from the visual, which was not real, to the essence, which was or might be. Mallarmé explicitly rejected 'the claim to enclose in language the material reality of things'.[1] In the same way painters such as Paul Klee and sculptors such as Henry Moore rejected representational art, and in music Schönberg abandoned the tonal centre and introduced the technique of atonality.

Behind this experimentation with form was a conscious and resolute determination to come to grips with 'the task of mastering reality afresh'.[2] It was the result of a crisis in standards and values which set in everywhere shortly after 1900. Schönberg published his revolutionary *Three Piano Pieces* (op. 11) in 1908. In Paris cubism emerged as a conscious and coherent movement in 1907. In Germany

1. ibid., p. 130.
2. cf. H. L. C. Jaffé, *Twentieth-Century Painting* (London, 1963), p. 12.

the formation of the group known as *Die Brücke* (1905), followed by the 'new secession' (1910) and *Der blaue Reiter* (1911), marked the birth of expressionism. All were impelled by the collapse of the old belief that positive truth was contained in sense perception and by the problems of reality revealed by science and technical discovery and the conclusions the human mind had drawn from them. By now, the question of the nature of reality had developed into the problem whether there was a reality at all that could be grasped, and how it could be grasped. The different styles, mannerisms, techniques that followed all represented different attempts to grapple with this problem. In France the Fauves, in Germany expressionists such as Heckel, Nolde, and Kandinsky used discordant colours, violent contrasts and brutal distortion to break through appearances to a truth 'truer than literal truth'. Stravinsky's early music employed similar methods for similar ends. But it was cubism which came closest to a new view of the world. It did so because it was more intellectual and less involved in 'an augmented emotional transcription of the artist's reaction than expressionism'.[1] For cubists the world existed, as it did not for the symbolists, but under the impact of new scientific theory it was conceived in a new way. In this they were like the artists of the Renaissance who also sought to assimilate art with the scientific discoveries of their day. Cubist painting was 'a research into the emergent nature of reality', an 'examination of reality in its many contingencies', 'an analysis of the multiple identity of objects'; it was 'painting conceived as related forms which are not determined by any reality external to those related forms'.[2] The universe the cubists depicted was one where things have no simple locations, and their rejection of a single viewpoint

---

1. cf. B. S. Myers, *Expressionism* (London, 1963), p. 43.
2. cf. Wylie Sypher's extremely able interpretation, *Rococo to Cubism in Art and Literature* (ed. New York, 1963), pp. 270, 276, 287, 288.

revealed a fuller vision of reality than was possible in any art based on the artificiality of Euclidean geometry. From this point of view cubist art had the sharpness and clarity of a scientific analysis.

'The ideas behind cubist painting', it has been said,[1] 'are reflected in all the modern arts.' Nevertheless, in spite of its impact, cubism was only a momentary halting place. For this there were two main reasons. The first was that the cubists themselves, by breaking up the object into its simplest elements – or, as in Picasso's famous *Man with Violin* (1911), resolving it into a series of planes – destroyed the object and opened the door for abstract art and a new wave of iconoclasm. The second was the impact of the First World War. It was no accident that dadaism and surrealism reached their peak between 1919 and 1921. The shock and disillusionment of war shook all faith in meaningful reality and gave point both to the bitter expressionist protest of Georg Grosz and Otto Dix and to the surrealist nightmares of Salvador Dali. 'Had they deceived us,' Eliot was later to ask[2] –

> Or deceived themselves, the quiet-voiced elders,
> Bequeathing us merely a receipt for deceit?

For the poets and artists who flocked into the modernist movements, the slaughter between 1914 and 1918, and the mock peace which crowned it, signified the bankruptcy not only of the existing order but of the system of values of a whole civilization. Judging them by their results, they had no further use for them. Hence surrealism took shape as the 'refusal of the modes of thought and feeling of traditional humanism'. But throughout the experimentation of the nineteen-twenties, side by side with the deliberate attempt to shock, the struggle to discover new ways of grasping reality was never abandoned. Mondriaan, in

1. ibid., p. 265.
2. In *East Coker* (1940); cf. *Collected Poems, 1909–1962* (London, 1963), p. 198.

particular, aimed in his abstract geometrical compositions at expressing 'pure reality', purged of 'the deadweight of the object'. The life of modern man, he wrote in 1917, was 'gradually being divorced from natural objects' and 'becoming more and more an abstract existence',[1] and he set himself the task of creating forms which would express this new situation. But the attempt to find a new insight into reality through abstraction did not survive the Second World War. In the post-Hitler, post-Hiroshima world Mondriaan's search for an harmonious balance was no longer acceptable. It seemed 'a dishonest extenuation',[2] and emphasis shifted from the attempt to picture reality to an existentialist attempt to express a new feeling about life. In the 'tachist' painting of Pollock or Appel what the scientist calls the 'field' became more important than the objects in the field. This painting expressed 'a world-view where the object disappears into patterns of behaviour': nothing, it has been said, 'could more effectively dismiss the romantic belief in freedom, individualism, and the importance of the decisive act'.[3]

3

Cubism, dadaism, and abstract painting, like the music of Bartok and Schönberg, or like Joyce's *Ulysses* and Kafka's *The Castle*, were entirely remote from the world of the nineteenth century. They are also entirely remote from the world of today. It is none of my business, even if I were competent to do so, to attempt their artistic evaluation; in the present context it is sufficient to see them as a transition from one stage of civilization to another. What happened, by the end of the transition, was the jettisoning

1. For a translation of Mondriaan's programmatic article from the first issue of the periodical *De Stijl*, cf. M. Seuphor, *Piet Mondrian. Life and Work* (London, 1957), pp. 142 ff.
2. Jaffé, op. cit., p. 26.
3. Sypher, op. cit., pp. 326–7.

of the inherited baggage of European culture. As Ortega wrote, the nineteenth century was bound to the past, on whose shoulders it thought it was standing; it regarded itself as the culmination of past ages. The present – he was writing in 1930 – recognized in nothing that was past any possible model or standard. The Renaissance revealed itself as 'a period of narrow provincialism, of futile gestures . . .'

We feel that we actual men have suddenly been left alone on the earth. . . . Any remains of the traditional spirit have evaporated. Models, norms, standards are no use to us. We have to solve our problems without any active collaboration of the past.[1]

It was this sense of alienation, of disinheritance, of the individual's incommunicable solitude, that was the framework of art and writing in the years before and after the First World War. The plays of Ibsen and Chekhov and novels such as Thomas Mann's *Buddenbrooks* (1901) had portrayed the crisis of the old society; they were, in a sense, its requiem. Rilke, above all others, emerged as the poet of a world from which doubt had dislodged all certainties; a world in which good does no good and evil no harm, in which lovers seek separation, not union, in which the whole accepted order of correspondences had collapsed like a house of cards.[2] Proust, seeking to perpetuate by an act of memory patterns and relationships which dissolved even as they were thought of, was its greatest novelist. Since there is no logical sequence, no casual development, since man is not a single unified person whose destiny is decided by his own actions or because the forces of nature are too overpowering for him to deal with – since he falls by the

1. cf. J. Ortega y Gasset, *The Revolt of the Masses* (ed. 1961), pp. 27–8. Ortega had already discussed this situation in his volume, *The Dehumanization of Art*, originally published in Madrid in 1925.
2. cf. Erich Heller, *The Disinherited Mind* (Penguin ed., 1961), p. 151.

wayside because life is meaningless or because he is a pur-
poseless bundle of atoms thrown haphazardly into the
dark emptiness of space – nothing remains but to communicate, seemingly at random, whatever the writer's
sensibility brings to the surface at the moment. The ultimate refinement – some would say the *reductio ad ab-
surdum* – was the surrealistic word sequences of James
Joyce, Gertrude Stein, and E. E. Cummings. A not unsympathetic writer said that Gertrude Stein, in using words
for pure purposes of suggestion, had 'gone so far that she
no longer even suggests'.[1] It was a criticism which applied
more generally. Of Schönberg similarly it was said that
his music had 'become so abstract, so individual and so
divorced from all relation to humanity as to be almost
unintelligible'.[2] Some of the greatest artists, sensing that
they were heading for a dead end, drew back. Stravinsky,
for example, recoiled after 1923 from his early 'dynamism'
to neo-classicism; Picasso quickly returned from his 'adventures on the borderline of the impossible'[3] and refused to
be bound by any single formula. But, in general, there
was an evident tendency for art to degenerate into a man-
nerism, and for artists and writers to break up into coteries
whose thoughts were too esoteric to strike a responsive
chord.

For the most part the experimentation which was
characteristic of the first half of the twentieth century
failed to arrive at positive results; certainly it failed to
produce a new synthesis. It would be a mistake to take this
failure too tragically. Many of the writers and artists of
the period were frankly destructive in purpose and had no
ambition to build anew; their object was simply to clear

1. cf. E. Wilson, *Axel's Castle. A Study in the Imaginative Literature of 1870–1930* (ed. London, 1961), p. 195.

2. cf. A. Einstein, *A Short History of Music* (6th ed., London,
1959), p. 201.

3. cf. E. H. Gombrich, *The Story of Art* (10th ed., London, 1960),
p. 435.

the ground and break with the past. The result nevertheless is that much of their work has retained only historical interest. This is true in the main, for example, of the bitter social comment of the German expressionists, and of writers such as Heinrich Mann and Ernst Toller, who were closely associated with them. For the rest, the attempts in the first thirty years of the twentieth century to make the necessary adjustments to the new world that was arising, never quite succeeded. It is an observation which has been made, for example, with specific reference to Rilke. 'His attempts to adjust himself to the new world', it has been said,[1] 'have a moving helplessness in a poem like the *Sonette an Orpheus*, an alienating helplessness in the *Elegien*.' In any case, what we note about his work today is its irrelevance to the contemporary world.

In this respect Rilke was in no way a solitary exception. Symbolism and expressionism also failed to maintain their hold. As Kafka said, symbols are 'of no use in daily life, which is the only life we have'; they 'merely express the fact that the incomprehensible is incomprehensible, and we knew that already'.[2] As for the revolt against the machine age, against the dreary desolation of *The Waste Land*, against the whole encroaching advance of standardized civilization, which was a recurrent theme from Baudelaire to Verhaeren and García Lorca, what was it but an inverted romanticism, a failure to come to terms with inescapable facts, a futile protest, comprehensible at the time but necessarily transient? Certainly it was

1. Guardini, op. cit., p. 126.
2. 'Concerning this', Kafka continued, 'a man once said:
  Why such reluctance? If you only followed the symbols you would become symbols yourselves, and thus rid of all your daily cares.
  Another said: I bet this is also a symbol.
  The first said: You have won.
  The second said: But unfortunately only symbolically.
  The first said: No, in reality; symbolically you have lost.'
(Heller, op. cit., pp. 188–9.)

foreign to a generation aware that industrialism had become the basis of the only society they would ever experience.

Equally marked was the failure of the prevailing artistic and literary modes to bridge the gap to the scientific and technical revolution which was the most distinctive feature of the age. This was not contradicted by the acute consciousness of the changing connotations of reality in Eliot's later writing – in *Burnt Norton* (1935), for example, or in *The Dry Salvages* (1941) – or in cubist and abstract painting. No one would deny that here and elsewhere there are overtones of modern scientific theory. But of the positive effects of natural science in changing the face of the world and the whole condition of our lives, little was assimilated. C. P. Snow noted the slowness of novelists, in America and elsewhere, in coming to terms with the facts of modern industrial society.[1] More fundamentally, no poet emerged capable of expressing the basic concepts of modern science, as Lucretius did with those of Democritus or Pope with Newtonian physics.

The position was summed up by a scientist writing in the early months of the Second World War, and looking back over the inter-war years. 'All the cultural activities of our epoch,' he wrote[2]

have failed in their main function. Neither painting nor literature has been able to arrive at a point of view positive and definite enough to be worth even considering as a basis for a new society. They have been very useful; they have cleared away a lot of mess which everyone wanted to see the last of; they have indicated, rather hazily, the direction in which a new outlook on the world might be found; but they have not drawn the curtains and enabled us to look through on to a promised land.

1. C. P. Snow, *The Two Cultures and the Scientific Revolution* (Cambridge, 1959), p. 29.
2. cf. C. H. Waddington, *The Scientific Attitude* (2nd. ed., 1948), p. 70.

4

After 1945 there was a considerable shift of ground. The preoccupations of the inter-war years, remote from those of the nineteenth century, were almost equally remote from those of the post-war world. It was not so much that they were rejected as that they were left behind. The incessant speculation of Eliot and Valéry as to what constitutes poetry, what function it performs, whether there is any point in writing it, no longer aroused the same response. Proust's introspective world of the private imagination had been so thoroughly explored and exploited that its possibilities seemed to have been exhausted. And the mood of pessimism and despair and resignation, of wandering helplessly without moorings in a world without hope, had worn thin. Perhaps its last great expression, written in 1952 under the shadow of the atom-bomb, was Beckett's *Waiting for Godot*, in which the two protagonists waited helplessly in a desolate world for the moment when 'all will vanish and we'll be alone again, in the midst of nothingness'.

*Waiting for Godot* was the end of the chapter which opened with Kafka's *The Castle*. Ten years later, when the hydrogen-bomb had become almost as familiar as the kitchen-table, people had learnt, or had begun to learn, to accept the incertitudes of the new world as a part of life. If the dominant intellectual personality of the decade after 1945 was Camus, whose message was the negative one of revolt, the typical figure in the following decade was Brecht, whose work presupposed a universe of relative values, where there were neither heroes nor saints, but in which human life had as its purpose to overcome the precarious and provisional state of human society. A similar attitude was reflected in the existentialist philosophy of Heidegger, Jaspers, and Sartre, the only widely influential philosophy of the period. Here also was an attempt to break away from the negative approach characteristic of the logical positivists. Existentialism, indeed, admitted no transcendental values; the individual was alone, but he was alone among others, involved in a situation which, although he had no hand in its creation, could not be avoided by escaping, like the symbolists, into a private inner world. The adjustment of Sartre's thinking, in the short span between *Huis clos* (1944) and *Les chemins de la liberté* (1945), can be taken as the measure of a basic change in human attitudes, a shift away from the standpoint of the isolated individual to one in which the individual is absorbed in a social reality that is intensified by the accelerating rhythm of new technical processes.[1]

Among the factors which brought about this change in point of view was the progress of sociology and the permeation of thought at all levels by notions derived from sociological investigation. Sociology taught that the group, not the individual, was the basic unit of society. It no longer started with the individual as the central concept in terms of which society must be explained, and it saw in group patterns of behaviour the norm which determined individual action. It is possible to question the social consequences of such views – they have been castigated in the United States by writers such as W. H. Whyte – but not their effectiveness. Their importance lay in the stimulus they gave to the shift, which was already taking place, from an egocentric and ultimately tedious preoccupation with the individual's personal fate and the malady of the European soul to the problems arising out of social relationships within the new, technical, industrial mass societies into which the changes of the last sixty or seventy years had plunged the world. The literature of protest and revolt – the characteristic product of an old order in decline – seemed to have shot its bolt, and a movement began away from subjectivity in the direction of

1. For the above, cf. M. Crouzet, *L'époque contemporaine. À la recherche d'une civilisation nouvelle* (2nd ed., Paris, 1959), pp. 452–3, 462–3, 466–7.

objectivity. In post-war Germany poets like Hans Magnus Enzensberger repudiated pure aesthetics and developed a lively social criticism.[1] Writers of the new generation – Robbe-Grillet for example – side-stepped the old labyrinth of introspection, seeking instead to show that the world 'quite simply *is*'.[2] Poets and artists who echoed Valéry's 'je ne suis curieux que de ma seule essence', ceased to be typical; and people turned instead to the question whether, in spite of its complexity and the strains it imposed, it was possible to come to terms with industrial society. They turned also to the question, thrust on them by social change and the spread of literacy, of the 'reintegration of art with the common life of society'.

5

The question of art and society, or of culture and civilization, was not new. It had been argued and debated ever since the Industrial Revolution.[3] But it acquired a new dimension when the results of the introduction of universal compulsory education, as it became general in the period after 1870, began to be felt. By the beginning of the nineteen-thirties the question had become the preoccupation of a generation. It had two sides: first, whether culture could survive in the new social environment, and, secondly, whether society could survive without the binding force of a 'common culture'.

For anyone looking back over the subsequent controversies the main impression left is one of sterility, and for that reason little would be gained by following their course in detail. The prevailing mood was largely pessimistic. Mass civilization, it was commonly asserted, was incompatible with culture. Culture was the work of

1. cf. Abosch, op. cit., p. 148.
2. cf. Sypher, op. cit., p. 329.
3. cf. Raymond Williams, *Culture and Society, 1780–1950* (Penguin ed., 1961).

minorities, and the domination of the masses, in conjunction with 'levelling', standardization and commercialization, implied the decline of civilization to the level of dull, mechanical uniformity. 'Civilization and culture', F. R. Leavis wrote in 1930, 'are coming to be antithetical terms,' and Yeats foresaw 'the ever increasing separation from the community as a whole of the cultivated classes'.[1] Writers and artists recoiled from the empty façade of urban life and the routine of mechanical civilization, believing with Yeats that the world of science and politics was somehow fatal to the poetic vision. Science, I. A. Richards declared, had deprived man of his spiritual heritage; a God who was subject to the theory of relativity could not be expected to provide inspiration for the poet.[2] But the main burden of complaint, expressed with particular force by T. S. Eliot, was that 'mass-culture' would always be a 'substitute-culture' and that in every mass society there was a 'steady influence which operates silently ... for the depression of standards'.[3]

This cultural pessimism, which reached its peak in A. J. Toynbee's *Study of History* – particularly in the ninth volume (1954) with its gloomy lament for the ills of western civilization – was comprehensible as a reaction against the complacent assumption, common among liberal-minded intellectuals at the beginning of the century, that the spread of literacy would automatically bring about the dissemination of the existing culture through the whole of society. There was never any reason why it should. The expectation that the new activated classes would simply absorb the literary, artistic, and moral standards of the old was contrary to all historical experience.

1. cf. F. R. Leavis, *Mass Civilization and Minority Culture* (Cambridge, 1930), p. 26; W. B. Yeats, *A Vision* (London, 1926), p. 214.

2. cf. I. A. Richards, *Science and Poetry* (London, 1926), p. 50.

3. cf. T. S. Eliot, *The Idea of a Christian Society* (London, 1939), p. 39; *Notes towards the Definition of Culture* (2nd ed., London, 1962), p. 107.

But the assumption that the breakdown of the prevailing scheme of values under the impact of social change was the same thing as the decline of all culture, was not very plausible either. It was easy to accuse the masses of indifference to serious literary and artistic activities and to blame them for the alleged gap between culture and civilization; but it was equally important to ask whether artists and writers had anything to say that was relevant to the new audience, or whether they had lost touch, in their idiom and values, with changing social realities. No one denied that there was a vast public (not necessarily of one class) devoted to trivial entertainment, commercialized art, escapist writing and cheap music; but this was not peculiar to modern society and its existence proved nothing. What was certain, on the other hand, was that the new public, which the spread of literacy had created on a world-wide scale, was different in its tastes and preoccupations from the fairly homogeneous educated *élite* to which writers and artists had hitherto addressed themselves. Its social background was far wider and the problems which attracted its attention were no longer those which had attracted attention in the minority culture of the past. When Jimmy Porter said contemptuously that he had written a poem 'soaked in the theology of Dante, with a good slosh of Eliot as well', he was speaking for a generation for which the rarified aesthetic values of the thirties were portentous bunk.

The emergence of literary and artistic forms capable of expressing the results of half a century of rapid social change was retarded – and is still in many respects retarded – by persistent attempts to salvage remnants of the old culture and graft them on to 'the new world of technological anonymity'. It was held back also by the dislocation and restrictions and frustration which typified the aftermath of the Second World War. But from about 1955 a breakthrough on a wide front was apparent. It was marked in England by Osborne's *Look Back in Anger*

(1956), in its way as characteristic an expression of a new social situation as *A Doll's House* had been in 1879 or *The Cherry Orchard* in 1904. It had already been expressed, almost a decade earlier, in the neo-realism of the Italian cinema. In 1958 it invaded the British cinema, which turned its back on the conventions of pre-war middle-class life and set out to investigate the social landscape of factory and public house, back-to-back cottages, and Saturday nights and Sunday mornings. It was not great art, but it was relevant art. It reflected a basic shift in class structure, and in the novels of Kingsley Amis or John Wain it turned in mockery and disgust against the bourgeois 'establishment' and the values it represented.

It was, perhaps, hardly accidental that the media which first sought to come to terms with the new realities were the cinema, the novel, and the drama. Architecture, under the inspiration of Frank Lloyd Wright and Walter Gropius, had already discovered forms of expression which were functionally adapted to a technological age; indeed, it might be held – in spite of the rapid commercialization and debasement of the new styles – that it was architecture which took the lead. Great projects of civil engineering such as the Rockefeller Center in New York, or the vast interlacing elevated motorways of cities such as Los Angeles, expressed with precision the 'spirit' – as well as the potentialities and limitations – of the new technological civilization. The classical forms of artistic expression – except perhaps music – experienced greater difficulty in bridging the gap and finding a new idiom. Poetry, in particular, with its intensely personal world, had difficulty in reaching a new audience; in western Europe at least, it seemed by the end of the Second World War as though it had exhausted its resources. But elsewhere – in Spanish America, for example – there were signs of a new beginning, and in Russia after Stalin a new phase commenced. As Isaac Deutscher has pointed out, whatever his literary merits, Pasternak spoke for a generation which was 'making its exit' and whose attitude towards life was not that of younger people: Yevtushenko represented the emergence of a new outlook on the world.[1]

These few indications of the breakthrough of new attitudes, inadequate though they are, suggest at least the basic nature of the change – namely, from a negative to a positive reaction and from rejection of technological civilization as incompatible with culture to acceptance of its challenge. This did not imply affirmation of the new society, in the sense in which affirmation was demanded of artists in the Soviet Union – with stultifying results – in the Stalinist period, but it did imply recognition of its inescapability. In this sense it is legitimate – provided we ignore the Stalinist overtones attached to the phrase – to speak of a return to social realism. It was accompanied by a change of idiom. The reason was not merely that there was no longer a public capable of appreciating the allusive verse of T. S. Eliot, to understand which even the select and sophisticated audience to which it was addressed had to be provided with a glossary. Rather it was because the content and style of life of modern society was no longer directly related to the old poetic methods, because the new generation saw, heard and associated differently from its predecessors. How, asked Brecht,[2] when immense innovations were being wrought on all sides, could artists hope to portray them, if they were limited to the old means of art? The result, without doubt – in this the pessimists who lamented the decline of the old values were right – was a revulsion against traditional humanism and the personality cult which lay at its core. The crisis through

1. cf. I. Deutscher, *The Great Contest* (London, 1960), p. 34; on Yevtushenko as the representative figure of a new 'creative wave' originating among the younger generation in 1957, cf. K. Mehnert, *The Anatomy of Soviet Man* (London, 1961), p. 168.
2. cf. E. Fischer, *The Necessity of Art* (Penguin ed., 1963), p. 114.

which society had passed, Romano Guardini pointed out,[1] was due at least in part to the fact that 'it received its historical imprint from the attitudes of a personality cult' which was no longer relevant. When the rise of technological civilization brought new social categories to the fore the preconditions changed. People were no longer prepared to accept unquestioningly the old assumption that 'the autonomous subject is the measure of human perfection', and culture, in the sense in which it had been understood through modern history, ceased to be regarded as 'a dependable rule for action'.[2]

By the end of the nineteenth century the impact of technology was changing the face of the world, but its effects on basic human attitudes were negligible. No creation of the human mind was more original than modern mathematics, but it by-passed the cultured classes and only a narrow circle of specialists troubled to learn its language. Even thirty years later European culture was still under the thrall of humanist traditions and literary values established under totally different conditions at the time of the Renaissance. The upheavals of the Second World War changed the whole situation; and in the succeeding decade a generation inspired by the potentialities of science and technology broke through the humanist barrier and took possession of the field. It was an irreversible victory. It implied the emergence of new criteria, linked to the great task, which science had set itself, of subduing nature and dominating the universe; and precisely because the demands this task made on humanity were so immense, they called for a new scale of values. Large-scale projects such as space programmes called for overall planning and a combination of skills which could only be achieved by teamwork – that is to say, if people were ready to accept a measure of discipline and conformity formerly rejected as incompatible with human dignity.

1. cf. Guardini, op. cit., p. 85.
2. ibid., pp. 76, 96.

The result was a new attitude to man's place in the universe. Naturally, the old intractable problem of the individual and his relations with the world around him was not disposed of – how could it ever be? – but it was put in a new context. In Hoyle's expanding universe the underlying anthropomorphism of the humanist tradition ceased to be credible, and with it the old personality cult. People no longer imbued liberty of external action or freedom of internal judgement with the same transcendent value as in the past, or aspired in the same way to live their lives according to principles uniquely their own. They knew that, in the complex, highly articulated society in which they lived, the old individualistic ethic no longer provided relevant standards and that solidarity and cooperation were at least as important. 'When all other substantial values have disintegrated,' Guardini said, these would remain as 'the supreme human values' in the new society which had emerged at the end of the long transition from modern to contemporary history.[1]

6

If we wish to measure the impact of the change in human attitudes I have attempted to sketch – a change consequential upon acceptance of the social implications of science and technology – it is important to realize that, like so much else in the contemporary world, it is not confined to Europe. Indeed, it might be true to say that the most significant aspect of the new outlook is its worldwide character. This is a consequence, in the last analysis, of the spread of industrialization, town-life, mass production and modern forms of communications, as a result of which the basic features of technological civilization, once regarded as characteristic of western Europe and North America, are rapidly becoming universal. The potential consequences of these changes, and at the same time of

1. ibid., pp. 73, 78, 84–5.

the spread of literacy, have as yet scarcely been grasped. Fifty years ago, the significant artistic and cultural movements radiated from Europe; today, as a result of the rapid spread of literacy and education, this has ceased to be the case. Already in the years between the First and Second World Wars, the United States had asserted a new pre-eminence in the English-speaking world, and it was American writers such as Faulkner and Hemingway who set the pace. More recently, this process of diversification has gone further, and we can see the beginning of significant literary movements in Latin America, Africa and elsewhere. In other fields the newly rising countries are already moving ahead. The great Mexican school of painting, represented by Siqueiros, Orozco and Rivera, had already made its impact before the Second World War; Japan has won a place of distinction in the art of the cinema; and in town-planning and architecture cities such as Rio de Janeiro, São Paulo, and Brasilia are unsurpassed anywhere in the world.

These developments are, of course, still at their beginning; but they are a sufficient indication that social change, however profound, may be a sign of renewal, not of collapse. As Alfred Weber has pointed out, it is simply not true, 'despite all ideas to the contrary', that the industrial worker in the United States or in England has been 'depersonalized', and the transformation, in barely more than a generation, of the Russian *muzhik* into a receptive, skilful, self-respecting industrial worker, with an immense appetite for literature, indicates the enormous human potentialities that lie to hand.[1] Today we have to reckon with a similar transformation of the working classes in China, Africa, Egypt, and elsewhere. What is clear is that, for none of these newly awakened peoples,

1. cf. Alfred Weber, *Abschied von der bisherigen Geschichte* (Hamburg, 1946), pp. 237, 239; there is an English translation under the title *Farewell to European History* (London, 1947), pp. 169, 170–1.

is the traditional culture – either their own or that of western Europe – a sufficient answer. Even in western Europe the old literary culture, with its intensely personal preoccupations, touched the life of ordinary people at too few points; their whole scale of values was based not on individual but on group activity – on the companionship of office and workshop, the inescapable closeness of the family unit, the enjoyment of leisure in the company of others – and a social ethic which idealized individuality bore little relation to the facts of their experience. The same is even more clearly true of the newly emancipated and literate workers elsewhere. Speculations and preoccupations of the type with which European writers and artists have tended to deal are alien to their experience; existentialism, with its anguish, its *néant* and its *nausée*, has little connexion, as a Mexican writer has pointed out, with American realities.[1] On these lines it would be illusory to expect a 'reintegration of art with the common life of society'.[2] But the changes outlined above indicate that a turning-point has been reached and that the gulf between cultural and social development, which had been growing wider ever since the Industrial Revolution, is again closing. With the new social awareness, the shift from the individual in isolation to the individual in society, and above all the change of viewpoint from 'We' and 'They' to 'Us', some of the most stubborn obstacles have been removed. At the same time, they have provided the basis for a civilization which, without losing its specific national and regional modes of expression, is truly universal in its connotations.

As late as 1959 C. P. Snow could still maintain that, whereas 'the scientists have the future in their bones', 'the traditional culture responds by wishing the future did not exist'.[3] When we consider how recently most of the changes which distinguish the contemporary world

1. cf. L. Zea, *América como consciencia* (Mexico, 1953), p. 160.
2. cf. Williams, op. cit., p. 286.     3. cf. Snow, op. cit., p. 11.

have become plainly visible, this time-lag should not be surprising. Consciousness and interpretation cannot precede creation. Eventually we may expect that art and literature will interpret the 'myth' of the contemporary age, and give expression to its beliefs and way of life. But just as their themes will be new, so also we must expect that they will reflect the change in the balance of world forces which is the clearest outcome of the events of the first half of the twentieth century. It is often argued that Europe, while losing its political hegemony, has retained and will continue to retain its cultural leadership; but this idea, though widely propagated, has little basis in fact. One of the most significant features of the contemporary age is the stimulus which the revolutions of the twentieth century, liberating them from their bonds to the past, to sterile forms and traditional themes, have given to the artistic and cultural life of other continents. Whereas by 1939 the poetry of most western countries was showing evident signs of exhaustion, new impulses were awakening in Asia, in Africa, and in Latin America. This evidence of cultural renewal on a world-wide scale is one of the most significant aspects of the contemporary scene.

It is only possible to give the barest indication of the scope and impact of the new cultural movements. They were usually associated – as the birth of European literature had been associated centuries earlier – with the vernacular; and they were influenced without exception by European forms and the stimulus of European ideas. This response to European influences – often but not exclusively to those of the French symbolists – has been variously assessed and interpreted, frequently in the sense that all that has emerged is a pale, imitative version of European models, severed from native tradition. Anyone aware of the degree to which European literature at the time of the Renaissance and earlier was dependent for its forms and subject-matter on classical models will hesitate before accepting this evaluation. In reality, as Sir

Hamilton Gibb has said of modern Arabic literature, 'the problem has very little to do with deliberate imitation of the west'.[1] This statement may be interpreted more generally. 'What the foreign examples did,' an American writer has said of Japan, 'was to afford the Japanese the means of expressing their new ideas and their consciousness of being men of the enlightened Meiji era'; but 'unless the Japanese had felt a need to create a new literature, no amount of foreign influence would have mattered'.[2] Even so, it is undeniably true, in all the countries that came under the impact of the west, that most early writing was derivative and of small intrinsic literary merit. This applied not only to the first stirrings of new currents in Arabic literature in the nineteenth century; it was true also of Japanese writing before the great period of creativity between 1905 and 1915, and of early Anglo-Ceylonese literature which moved 'with occasional pleasant detours through an imitation of English models to an inevitable dead end'.[3] But these, as Sir Hamilton Gibb has written,[4] were the 'precursors', and their importance lies less in what they achieved than in the influence they exerted and the new currents they set in motion. Before long a new stage was reached. It occurred in South America, for example, around 1925, when the search began 'for an artistic expression that should be our own and not subservient to Europe';[5] and in Japan the publication of Nagai Kafu's *The River Sumida* in 1909 has been picked out as marking the transition 'from a period when European works were slavishly imitated to one when an awareness and recep-

1. cf. H. A. R. Gibb, *Studies on the Civilization of Islam* (London, 1962), p. 298.

2. cf. D. Keene, *Modern Japanese Literature* (London, 1956), p. 16.

3. cf. *Ceylonese Writing. Some perspectives*, ed. C. R. Hensman (*Community*, vol. 5, Colombo, 1963), p. 67.

4. op. cit., pp. 258, 286, 292.

5. cf. P. Henriquez-Ureña, *Literary Currents in Hispanic America* (Cambridge, Mass., 1945), p. 192.

tivity to them was not permitted to blot out the native heritage'.[1]

What the new literary movements of Asia acquired from the west – including Russia – was, above all else, a model for a flexible idiom, through which to express the thoughts and ideals of modern civilization. It was here, in the liberation from obsolete images and rigid conventions, that the western impact was strongest. Everywhere in the east the old literary style – involved, periphrastic, and obscure – was out of touch with present reality; it was the creation of a small *élite*, a mystery in which only the scholastically educated might participate, but above all else it lacked the resources to express the thought and ideals of modern society, walling them off instead in a separate compartment dividing art from life.[2] Even the syntax required to be adjusted to meet modern methods of reasoning and feeling. The Egyptian writer, Husayn Haykal, for example, explained bitterly in 1927 his feeling of rebellion when he was unable to express in his own language what he felt in his heart and instead found the appropriate English or French expressions forming in his mind.[3] The result everywhere was a revolt – resisted by conservatives and traditionalists – against the old literary forms and, particularly where, as in China or the Moslem world, the literary language was no longer the language of everyday speech, a deliberate use of the colloquial or vernacular idiom as the only appropriate vehicle for fresh ideas. This turning to the vernacular began early in Egypt and was continued by writers elsewhere in the Arab world, such as the Iraqi, Abd al-Malek Nouri.[4] But it was in China that the issues were most clearly formulated.

1. Keene, op. cit., p. 25.
2. cf. J. R. Levenson, *Confucian China and its Modern Fate*, vol. 1 (London, 1958), p. 127.    3. cf. Gibb, op. cit., p. 274.
4. For Egypt, cf. Gibb, op. cit., pp. 254, 272, 294, 299; for Abd al-Malek Nouri and the post-war writers of Iraq – 'tous . . . hantés par les problèmes de l'actualité politique et sociale' – cf. *Orient*, no. 4 (1957), p. 18.

Here the Hsin ching-nien group, the 'new youth' who gathered round Chen Tu-hsiu and Hu Shih in 1916 and 1917, inaugurated a cultural revolution the importance of which it would be hard to exaggerate. One writer, indeed, has gone so far as to suggest that, for the historian of the future, it may turn out to be a more significant event in Chinese history than many of the political revolutions in which historians have sought the clues to recent developments.[1]

The literary revolution in China epitomizes the changes which underlie the cultural revival in the extra-European world.[2] The essential point is that literary reform was part of the national awakening; indeed, it could be said that it was the most important part because, as Chen Tu-hsiu wrote, 'purely political revolution' – since it brought no change in the fields of ethics, morality, literature and the fine arts – was 'incapable of changing our society'. Hu Shih denounced literary Chinese as a dead language because it was 'no longer spoken by the people'. It had decayed because of 'over-emphasis on style at the expense of spirit and reality'. Furthermore the theory on which classical Chinese literature was based – namely, that its purpose was 'to convey the tao' (i.e. moral principles) – was too restrictive. Huang Yuan-yung had already set out that what was necessary was 'to bring Chinese thought into direct contact with the contemporary thought of the world' in order to 'accelerate its radical awakening'; and, he added, 'we must see to it that the basic ideals of world thought must be related to the life of the average man'. Hence the emphasis on the vernacular as the medium for creating a living literature, and hence also the three aims of the literary revolution which Chen formulated as follows:

1. K. M. Pannikar, *Asia and Western Dominance* (London, 1953), p. 504.
2. For the following, cf. Chow Tse-tsung, *The May Fourth Movement* (Cambridge, Mass., 1960), pp. 271 ff.

1. To overthrow the painted, powdered and obsequious literature of the aristocratic few, and to create the plain, simple and expressive literature of the people.

2. To overthrow the stereotyped and over-ornamental literature of classicism and to create the fresh and sincere literature of realism.

3. To overthrow the pedantic, unintelligible, and obscurantist literature of the hermit and recluse, and to create the plain-speaking and popular literature of society in general.

The years 1918 and 1919, the years of revolutionary ferment which found an outlet in the Fourth of May movement, were the time when Chen's principles were put into effect. 'After 1919', Hu Shih was later to write, 'vernacular literature spread as though it wore seven-league boots.' And with it spread a new social consciousness and a new attitude to China's problems. Its effects were reinforced after 1921 by the work of a new organization, the Society for Literary Studies, which undertook on a major scale the translation of western writing, particularly the literature of the 'oppressed peoples'.[1] The result was the collapse of the archaic literary language and the old stereotyped literary forms. Henceforward creative writing in China was modelled upon the literature of the west and had little or no connexion with the Chinese classics. But the significant effects of the western impact were not literary but social. The new horizons revealed by western literature, the comparisons which it made possible, were a potent factor in opening the eyes of the new generation to the realities of the Chinese social scene. By 1925 the early tendency, which contact with western

1. ibid., p. 285. More than twenty countries were represented, including Germany, France, Great Britain, the United States, Russia, Sweden, Spain, Norway, Austria, Hungary, Poland, Belgium, Israel, Holland, Italy, and Bolivia. French writers included Barbusse, Baudelaire, Anatole France, Maupassant, and Zola; the Scandinavians were Bjornson, Bojer, Ibsen, and Strindberg; and Russians included Andreyev, Artzybashev, Dostoyevsky, Gogol, Tolstoy, Turgenev, and Gorky. The Indian, Rabindranath Tagore, was also translated.

literature had stimulated, to individualism, pessimism, the expression of personal feeling and 'art for art's sake' was collapsing, the social aspects of literature were in the ascendant, the prevalent mood, fostered by the new literary figure, Mao Tun (b. 1896), was against aestheticism and towards realism. The spirit of the age, Mao announced, impelled the writer to the search for social truth; the thoughts and feelings which he expressed 'must be common to the masses, common to the whole of mankind, and not just to the writer himself'.[1]

In all these respects the course of development in China was representative of what was happening elsewhere. European influences provided the original literary stimulus; but very soon the impact of national, social, and religious movements in the countries concerned transformed the new literature from a derivative literary mode into a vehicle for expressing a new social situation. This was the case, for example, with the Tamil novel in Ceylon.[2] What western models provided was not content but 'vigour of thought and congruence with the present'.[3] In India, as Pannikar has said,[4] it was 'not Europe' but 'the New Life' that was echoed in the new writing – in 'poetry and prose in which the conditions of our existence are constantly related to the extreme limit of possibilities'. A similar development had already occurred in Spanish America.[5] Here the *modernistas*, disciples of the French symbolists, who withdrew from politics and devoted themselves to 'pure' literature, were superseded between 1918 and 1922 by writers and painters who strove to relate their art to the social movements of their countries. With poets such as the Chilean, Pablo Neruda (b. 1904), and

1. cf. Levenson, op. cit., p. 128; Chow Tse-tsung, op. cit., p. 284.
2. cf. Hensman, op. cit., p. 103.
3. Gibb, op. cit., p. 260.
4. op. cit., p. 505.
5. For the following cf. Henríquez-Ureña, op. cit., pp. 168–73, 185–98.

the Mexican, Octavio Paz (b. 1914), Spanish-American writing asserted its independent status. From about 1920 it was occupied with specifically Spanish-American problems, the fight with the jungle, the stresses and clashes of conditions unknown in Europe, and particularly the social problems of the negro and the Indian – 'Indio que labras con fatiga tierras que de otros dueños son,' in the words of the Peruvian poet, José Santos Chocano. From around 1925 a poetry of negro life, destined to have echoes in Africa, appeared in Puerto Rico and Cuba, a poetry of immense beauty which – for example, in the works of Nicolás Guillén – expressed the dilemma of the black man condemned to a permanent state of exile, lacking a tribal name, lacking a tolerated religion, lacking a recognized culture of his own and without power or influence in the new mixed societies that had been built upon his labour.

It was not only Spanish-American literature which was remarkable, as it threw off its dependence on European models, for its social realism, its intense concern with the desperate problems of society. We find the same characteristic, utterly distinct from the prevalent mood of western writing in the 1930s, at the same period in China and Japan, and in India it became dominant through the influence of the Progressive Writers' Association, founded in 1935.[1] In the Arab world the Egyptian modernists were inspired by the conviction that 'a literary revival, reflecting a revolution in the ideas and outlook of the people, is a necessary preliminary to a full revival of national life'. Their aim, in the words of Abbas Mahmud al-Aqqad, was not to create an intellectual culture, 'a culture of decadence and mere words', but a natural culture, 'a culture of progress'.[2] And, finally, the same profound social and political commitment was a potent element in the poetry

1. cf. Keene, op. cit., p. 27; Chow Tse-tsung, op. cit., p. 287; Pannikar, op. cit., p. 505.

2. cf. Gibb, op. cit., pp. 282, 286.

of Africa – in Sartre's estimation 'the true revolutionary poetry of our time' – as it awakened under the impact of the West Indians, Aimé Césaire and Léon Damas, and found full expression in Léopold Senghor's epoch-making volume of 1948, the *Anthologie de la nouvelle poésie nègre et malgache*.[1]

Here also, particularly in French Africa, we are brought face to face with the pattern already familiar in Asia and north of the Sahara – the use of forms and images of European origin to express ideas and sentiments profoundly un-European and often anti-European in content. Technically writers like Senghor or the Madagascan poet Rabéarivelo stand in the tradition of French symbolism, just as Césaire makes superb use of the surrealist idiom; but behind a similarity of form was a different experience – the great experience of an historic awakening at a time when European poetry was haunted by the disintegration of the old way of life. The themes of Senghor and his fellows were 'the bitter taste of liberty', 'the beating of the dark pulse of Africa in the mist of lost villages'.[2] The European cultural revolution, in short, was met by an African cultural counter-revolution, a rediscovery and reassertion of African values, expressed by poets such as David Diop in the theme of *négritude* –

> Souffre, pauvre Nègre,
> Nègre noir comme la Misère,[3]

– and its contempt for the white world –

> Listen to the white world
> how it resents its great efforts
> how their protest is broken under the rigid stars

1. Senghor's anthology is not easily obtainable. For a more recent and more fully representative selection in English, cf. Gerald Moore and Ulli Beier, *Modern Poetry from Africa* (Penguin Books, 1963).

2. Moore and Beier, op. cit., pp. 44, 58.

3. Quoted by T. Hodgkin, *Nationalism in Colonial Africa* (London, 1956), p. 75.

how their steel blue speed is paralysed in the mystery of the
flesh.
Listen how their defeats sound from their victories.
Listen to the lamentable stumbling in the great alibis.
Mercy! mercy for our omniscient naïve conquerors.
Hurrah for those who never invented anything
hurrah for those who never explored anything
hurrah for those who never conquered anything
hurrah for joy
hurrah for love
hurrah for the pain of incarnate tears.[1]

The themes of *négritude* and protest are not, of course,
the sole content of African writing.[2] The problem of
Africa, as the Ghanaian poet, Michael Dei-Aneng, has
pointed out, is that of a continent 'poised between two
civilizations', and admits of no simple answer. But its
impact is such that writers and artists of every category –
'believers and atheists, Christians, Moslems, and Communists alike', as Alioun Diop once expressed it – 'are all
more or less committed'.[3] This is the distinctive feature
of the resurgent literature of Asia and South America, as
well as of Africa; their writers and artists are engaged in
bringing out all that they can contribute to the building
up of a new civilization. They know that what they are
expressing is not the feeling of the people as a whole, but
the views of a minority who are striving to interpret the
events of the present for the people; but it is precisely in
this that they see their contribution to the future.

Taken as a whole, the literary evidence shows remarkable consistency. In the Far East and the Middle East,

1. Quoted by Colin Legum, *Pan-Africanism* (London, 1962), p. 93,
from a translation of Césaire's *Cahier d'un retour au pays natal*.

2. This is emphasized by Moore and Beier, op. cit., pp. 23–4, who
argue that 'the wellspring of *négritude* is running dry'. Certainly
the attitude of the poets of English-speaking Africa is different. It is
unnecessary to pursue the question here; the problems are discussed
by Legum, op. cit., pp. 94–103.

3. ibid., pp. 98–9.

north of the Sahara and south of the Sahara, on the Amazon and the Rio de la Plata and in the lands of the Andes, it brings before us new peoples arising, new energies seeking expression, a definite view of life set in conscious counterpoint to that of Europe. African poetry is strikingly free from the element of *Kulturmüdigkeit*, or weary disillusion, which settled like a blight on the previous generation of European writing. If, as I have suggested, that phase is now a thing of the past in Europe also, if European literature and art have made their peace with the new civilization of machinery and technology and the 'common man', if the mood of rejection and resignation has been followed by a mood of affirmation and the exploration of the new potentialities which science has opened up, then it may not be illusory to look forward to the synthesis which still eludes us – the 'development of a common culture' and 'the re-integration of art with the common life of society'. But its basis, and the experience it reflects, will be far wider than ever before; the answers will be given 'by humanity as a whole – not one country or one city, as in the past'.[1] Here the literary and artistic evidence is unequivocal. The European age – the age which extended from 1498 to 1947 – is over, and with it the predominance of the old European scale of values. Literature, like politics, has broken through its European bonds, and the civilization of the future, whose genesis I have tried in the preceding pages to trace, is taking shape as a world civilization in which all the continents will play their part.

1. cf. Jaffé, op. cit., p. 27.

# INDEX